Cornell University Library

BOUGHT WITH THE INCOME
FROM THE
SAGE ENDOWMENT FUND
THE GIFT OF
Henry W. Sage
1891

ENGINEERING LIBRARY

A.113609   14/4/1898

Cornell University Library
QE 262.A3D15 1886

The geology of the country around Aldbor

3 1924 004 543 447

[*All Rights Reserved.*]

MEMOIRS OF THE GEOLOGICAL SURVEY.

ENGLAND AND WALES.

THE GEOLOGY OF

THE COUNTRY AROUND

ALDBOROUGH, FRAMLINGHAM, ORFORD, AND WOODBRIDGE.

(EXPLANATION OF QUARTER-SHEETS 49 S. AND 50 S.E.)

BY

W. H. DALTON, F.G.S.,

EDITED (WITH SOME ADDITIONS) BY

W. WHITAKER, B.A., F.G.S., Assoc. Inst. C.E.

PUBLISHED BY ORDER OF THE LORDS COMMISSIONERS OF HER MAJESTY'S TREASURY.

LONDON:
PRINTED FOR HER MAJESTY'S STATIONERY OFFICE.

PUBLISHED BY
EYRE AND SPOTTISWOODE, EAST HARDING STREET, LONDON, E.C.
ADAM AND CHARLES BLACK, NORTH BRIDGE, EDINBURGH.
HODGES, FIGGIS, & Co., 104, GRAFTON STREET, DUBLIN.

1886.

Price One Shilling.

LIST OF GEOLOGICAL MAPS, SECTIONS, AND PUBLICATIONS OF THE GEOLOGICAL SURVEY.

THE Maps are those of the Ordnance Survey, geologically coloured by the Geological Survey of the United Kingdom under the Superintendence of ARCH. GEIKIE, LL.D., F.R.S., Director General.
(For Maps, Sections, and Memoirs illustrating Scotland, Ireland, and the West Indies, and for full particulars of all publications, see "Catalogue." Price 1s.)

ENGLAND AND WALES.—(Scale one-inch to a mile.)

Maps marked * are also published as Drift Maps. Those marked † are published only as Drift Maps.

Sheets 3*, 5, 6*, 7*, 8, 9, 11 to 22, 25, 26, 30, 31, 33 to 37, 40, 41, 44, 47*, 64*, price 8s. 6d. each.
Sheet 4, 5s. Sheets 2*, 10, 23, 24, 27 to 29, 32, 38, 39. 53, 84†, 85†, 4s. each.
Sheets divided into quarters; all at 3s. each quarter-sheet, excepting those in brackets, which are 1s. 6d. each.
1*. 42, 43, 45, 46, NW, SW, NE*, SE, 48, NW†, SW*, NE†, (SE*), (40†), 50†, 51*, 52 to 57. (57 NW), 59 to 63, 65†, 66 SW†, NE†, NW*, SE†, 67 N†, (S†), 68 E†, (NW*), SW†, 69†, 70*, 71 to 75, 76 (N) S, (77 N), 78, 79. NW*, SW NE*, SE*, 80 NW*, SW*, NE, SE, 81 NW*, SW, NE, SE, 82, 83*, 87, 88, NW NE, SE, 89 NW*, SW*, NE, SE*, 90 (NE*), (SE*), 91, (NW*), (SW*), NE*, SE*, 92 SW, SE, 93 NW, SW, NE*, SE*, 94 NW†, SW†, (NE†), SE†, 95 NW*, NE*, (SE*), 96*, 97 SE, 98, 99 (NE*), (SE*), 101 SE, 102 NE*, 103*, 104*, 105 NW, NE*, SE, 106 NE* SE*, 109 SW, SE*, 110 (NW*), (NE*) SW*.

HORIZONTAL SECTIONS, 1 to 139. England, price 5s. each.   **VERTICAL SECTIONS,** 1 to 75, England, price 3s. 6d. each.

COMPLETED COUNTIES OF ENGLAND AND WALES, on a Scale of one-inch to a Mile.

Sheets marked * have Descriptive Memoirs.   Sheets or Counties marked † are illustrated by General Memoirs.

ANGLESEY†,—77 N, 78. Hor. Sect. 40.
BEDFORDSHIRE,—46 NW, NE, SW†, SE†, 52 NW, NE, SW, SE.
BERKSHIRE,—7*, 8†, 12*, 13*, 34*, 45 SW*. Hor. Sect. 59, 71, 72, 80.
BRECKNOCKSHIRE,—36, 41, 42, 56 NW, SW, 57 NE, SE. Hor. Sect. 4, 5, 6, 11, and Vert. Sect. 4 and 10.
BUCKINGHAMSHIRE,—7* 13* 45* NE, SE. 46 NW, SW†, 52 SW. Hor. Sect. 74, 79.
CAERMARTHENSHIRE†, 37, 38, 40, 41, 42 NW, SW, 56 SW, 57 SW, SE. Hor. Sect. 2-4, 7, 8, ; and Vert. Sect. 3-6, 13, 14.
CAERNARVONSHIRE†,—74 NW, 75, 76, 77 N, 78, 79 NW, SW. Hor. Sect. 28, 31, 40.
CARDIGANSHIRE†,—40, 41, 56 NW, 57, 58, 59 SE, 60 SW. Hor. Sect. 4, 5, 6.
CHESHIRE,—73 NE, NW, 79 NE, SE, 80, 81 NW*, SW*, 88 SW. Hor. Sect. 18, 43, 44, 60, 64, 65, 67, 70.
CORNWALL†,—24†, 25†, 26†, 29†, 30†, 31†, 32†, & 33†.
DENBIGH†,—73 NW, 74, 75 NE, 78 NE, SE, 79 NW, SW, SE, 80 SW. Hor. Sect. 31, 35, 38, 39, 43, 44; and Vert. Sect. 24.
DERBYSHIRE†,—62 NE, 63 NW, 71 NW, SW, SE, 72 NE, SE, 81, 82, 88 SW, SE. Hor. Sect. 18, 46, 60, 61, 69, 70.
DEVONSHIRE†,—20†, 21†, 22†, 23†, 24†, 25†, 26†, & 27†. Hor. Sect. 19.
DORSETSHIRE,—15, 16, 17, 18, 21, 22. Hor. Sect. 19, 20, 21, 22, 56. Vert. Sect. 22.
ESSEX,—1*, 2*, 47*, 48. Hor. Sect. 84, 120.
FLINTSHIRE†,—74 NE, 79. Hor. Sect. 43.
GLAMORGANSHIRE†,—20, 36, 37, 41, & 42 SE, SW. Hor. Sect. 7, 8, 9, 10, 11; Vert. Sect. 2, 4, 5, 6, 7, 9, 10, 47.
GLOUCESTERSHIRE,—19, 34*, 35, 43 NE, SW, SE, 44*. Hor. Sect. 12 to 15, 59; Vert. Sect. 7, 11, 15, 46 to 51.
HAMPSHIRE,—8†, 9†, 10*, 11†, 12*, 14, 15, 16. Hor. Sect. 80.
HEREFORDSHIRE,—42 NE, SE, 43, 55, 56 NE, SE. Hor. Sect. 5, 13, 27, 30, 34; and Vert. Sect. 15.
HERTFORDSHIRE,—1† NW, 7*, 46, 47*. Hor. Sect. 79, 120, 121.
HUNTINGDON,—51 NW, 52 NW, NE, SW, 64*, 65.
KENT†.—1† SW & SE, 2†, 3†, 4*, 6†. Hor. Sect. 77 and 78.
LANCASHIRE,—79 NE, 80 NW*, NE, 81 NW, 83 NW, SW†, 89, 90, 91, 92 SW, 98. Hor. Sect. 62 to 68, 85 to 87. Vert. Sect. 27, 34, 61.
LEICESTERSHIRE,—53 NE, 62 NE, 63*, 64*, 70*, 71 SE, SW. Hor. Sect. 46, 48, 49, 52, 122, 124, 125.
MERIONETHSHIRE†,—59 NE, SE, 60 NW, 74. 75 NE, SE. Hor. Sect. 26, 28, 29, 31, 32, 35, 37, 38, 39.
MIDDLESEX†,—1† NW, SW, 7*, 8†. Hor. Sect. 79.
MONMOUTHSHIRE,—35, 36, 42 SE, NE, 43 NW. Hor. Sect. 5 and 12; and Vert. Sect. 8, 9, 10, 12.
MONTGOMERYSHIRE†,—56 NW, 59 NE, SE, 60, 74 SW, SE. Hor. Sect. 26, 27, 29, 30, 32, 34, 35, 36, 38.
NORTHAMPTONSHIRE,—64, 45 NW, NE, 46 NW, 52 NW, NE, SW, 53 NE, SW, & SE, 63 SE, 64.
NOTTINGHAM,—70*, 71* NE, SE, NW, 82 NE*, SE*, SW, 83, 86, 87* SW. Hor. Sect. 60, 61.
OXFORDSHIRE,—7*, 13*, 34*, 44*, 45*, 53 SE*, SW. Hor. Sect. 71, 72, 81, 82.
PEMBROKESHIRE†,—38, 39, 40, 41, 58. Hor. Sect. 1 and 2; and Vert. Sect. 12 and 13.
RADNORSHIRE,—42 NW, NE, 56, 60 SW, SE. Hor. Sect. 5, 6, 27.
RUTLANDSHIRE†,—this county is wholly included within Sheet 64.*
SHROPSHIRE,—55 NW, NE, 56 NE, 60 NE, SE, 61, 62 NW, 73, 74 NE, SE. Hor. Sect. 24, 25, 30, 33, 34, 36, 41, 44, 45, 53, 54, 58; and Vert. Sect. 23, 24.
SOMERSETSHIRE,—18, 19, 20, 21, 27, 35. Hor. Sect. 15, 16, 17, 20, 21, 22; and Vert. Sect. 12, 46, 47, 48, 49, 50, 51.
STAFFORDSHIRE,—54 NW, 55 NE, 61 NE, SE, 62, 63 NW, 71 SW, 72, 73 NE, SE, 81 SE, SW. Hor. Sect. 18, 23, 24, 25, 41, 42, 45, 49, 54, 57, 51, 60; and Vert. Sect. 16, 17, 18, 19, 20, 21, 23, 26.
SUFFOLK,—47,* 48,* 49, 50, 51, 66 SE*, 67.
SURREY,—1 SW†, 6†, 7*, 8†, 12†. Hor. Sect. 74, 75, 76, and 79.
SUSSEX,—4*, 5†, 6†, 8†, 9†, 11†. Hor. Sect. 75, 76, 77, 78.
WARWICKSHIRE,—44*, 45 NW, 53*, 54, 62 NE, SW, SE, 63 NW, SW, SE. Hor. Sect. 23, 48 to 51; Vert. Sect. 21.
WILTSHIRE,—12*, 13*, 14, 15, 18, 19, 34*, and 35. Hor. Sect. 15 and 59.
WORCESTERSHIRE.—43 NE, 44*, 54, 55, 62 SW, SE, 61 SE. Hor. Sect. 13, 23, 25, 50, 59, and Vert. Sect. 15.

GENERAL MEMOIRS OF THE GEOLOGICAL SURVEY.

REPORT on CORNWALL, DEVON, and WEST SOMERSET. By Sir H. T. DE LA BECHE. 14s. (O.P.)
FIGURES and DESCRIPTIONS of the PALÆOZOIC FOSSILS in the above Counties. By PROF. PHILLIPS. (O.P.)
The MEMOIRS of the GEOLOGICAL SURVEY of GREAT BRITAIN. Vol. I., 21s.; Vol. II. (in 2 Parts), 42s.
NORTH WALES. By SIR A. C. RAMSAY. Appendix, by J. W. SALTER and R. ETHERIDGE. 2nd Ed. 21s. (Vol. III. of Memoirs, &c.)
The LONDON BASIN. Part I. Chalk and Eocene Beds of S. and W. Tracts. By W. WHITAKER. 13s. (Vol. IV. of Memoirs, &c.)
Guide to the GEOLOGY of LONDON and the NEIGHBOURHOOD. By W. WHITAKER. 4th Ed.   1s.

[All Rights Reserved.]

MEMOIRS OF THE GEOLOGICAL SURVEY.

ENGLAND AND WALES.

THE GEOLOGY OF

THE COUNTRY AROUND

ALDBOROUGH, FRAMLINGHAM, ORFORD, AND WOODBRIDGE.

(EXPLANATION OF QUARTER-SHEETS 49 S. AND 50 S.E.)

BY

W. H. DALTON, F.G.S.,

EDITED (WITH SOME ADDITIONS) BY

W. WHITAKER, B.A., F.G.S., Assoc. Inst. C.E.

PUBLISHED BY ORDER OF THE LORDS COMMISSIONERS OF HER MAJESTY'S TREASURY.

LONDON:
PRINTED FOR HER MAJESTY'S STATIONERY OFFICE.

PUBLISHED BY
EYRE AND SPOTTISWOODE, EAST HARDING STREET, LONDON, E.C.
ADAM AND CHARLES BLACK, NORTH BRIDGE, EDINBURGH.
HODGES, FIGGIS, & Co., 104, GRAFTON STREET, DUBLIN.
1886.

Price One Shilling.

PREFACE.

The area described in the present Memoir lies in the south-eastern part of Suffolk, and is shown in the two Quarter-sheets 49 S. and 50 S.E. of the Geological Survey Map of England and Wales. It includes nearly the whole of the Coralline Crag. Owing to the extent and thickness of the superficial deposits only one edition of the maps is issued, representing the distribution and variety of these deposits with such small tracts of the older formations as appear at the surface—the Chalk near Earl Soham, and the London Clay at the bottoms of some of the valleys in the southern part of the area. By the help of well-sections, however, the northward extension of the older Tertiary groups has been traced.

A full account of the literature devoted to the Red Crag will be found in the Memoir on the Geology of Ipswich; only those works that treat specially of the Coralline Crag are enumerated in the following pages. A general Monograph of the whole of the Pliocene formation of England is now in preparation.

ARCH. GEIKIE,
Director-General.

Geological Survey Office,
  28, Jermyn Street,
    1st November 1886.

E 18861. Wt. 10025.

NOTICE.

THE mapping of Sheet 50 S.E., and of the adjacent strip of sea-coast comprised in 49 S., was almost entirely done (under the superintendence of Mr. W. Whitaker) by Mr. W. H. Dalton, who also wrote the Memoir before his retirement from the Geological Survey.

In preparing the MS. for publication, Mr. Whitaker made some insertions from his own field-notes, and from published Memoirs by various Authors. He has also added the remarks on the Fossils of the Coralline Crag, on the literature of the subject, and on Shingle, besides contributing the larger part of the Appendix of Well-sections.

H. W. BRISTOW,
Senior Director.

Geological Survey Office,
28, Jermyn Street, S.W.
20th October 1886.

CONTENTS.

	PAGE
Preface by the Director General	iii
Notice by the Director	iv

CHAP. I. Introduction. Cretaceous and Eocene Beds.—Area. Rivers (by W. W. and W. H. D.) Physical Features. Geological Formations. Chalk. Reading Beds. London Clay - 1

CHAP. II. Coralline Crag. General Description. Fossils (by W. W.) Literature (by W. W.). Details - 5

CHAP. III. Red Crag. General Description. Details. (Valley of the Finn. Valley of the Deben, up the Right Side. Valley of the Deben, down the Left Side. Valley of the Butley River) - 12

CHAP. IV. Red Crag (continued). Details. (Valley of the Ore or Alde, up the Right Side. Valley of the Ore. Valley of the Alde and Saxmundham Valley. Valley of the Ore (or Alde), down the Left Side below Snape. Sea-board. Hundred River Valley and Minsmere Valley). Chillesford Clay - 20

CHAP. V. Glacial Drift. Divisions. Lower Boulder Clay. (Valley of the Deben. Valley of the Butley River. Valley of the Alde or Ore. Northward from Aldborough). Sand and Gravel. (Valley of the Finn. Valley of the Deben, up the Right Side. Valley of the Deben, down the Left Side. Valley of the Butley River. Valley of the Ore, up the Right Side. Valley of the Ore, down the Left Side, including the Valley of the Alde. Valley of The Hundred River. Valley of the Leiston Brook. Minsmere Valley) - 27

CHAP. VI. Glacial Drift (continued).—Upper Boulder Clay. (General Account. Valley of the Finn. Valley of the Deben, up the Right Side. Valley of the Deben, down the Left Side. Valley of the Butley River. Valley of the Ore, up the Right Side. Valley of the Ore, down the Left Side. Valley of the Alde. Saxmundham Valley. Valley of the Ore or Alde. Left Side. Valley of the Hundred River. Minsmere Valley. Valley of the Waveney) - 36

CHAP. VII. Post Glacial Beds (by W. W. and W. H. D.)—River Gravels. Alluvium. Coast Deposits (Shingle. Blown Sand) - 46

APPENDIX.—Well-sections. (By W. W. and W. H. D.) - 50

INDEX - 58

ILLUSTRATIONS.

	PAGE
Figs. 1, 2.—Casts of *Voluta Lamberti*, showing the Filling in of the Whorls of the Shells	11
„ 3.—Section in a Pit, three furlongs East of Great Bealings Church. (Wood.)	12
„ 4.—Section in a Pit on the Common about half a mile West of Butley Abbey. (Prestwich.)	13
„ 5.—Section in a Pit by the side of the Road $1\frac{1}{4}$ (? $\frac{3}{4}$) mile N.N.E. from Sudbourn Church. (Prestwich.)	20
„ 6.—Section in Ballast-pit, Aldborough. (Prestwich.)	23
„ 7.—Section in a Pit half a mile eastward of Blaxhall	29
„ 8, 9.—Views from the Top of Orfordness High Lighthouse. (Redman.)	48

THE GEOLOGY OF

THE COUNTRY AROUND

ALDBOROUGH, FRAMLINGHAM, ORFORD, AND WOODBRIDGE.

CHAPTER I.—INTRODUCTION. CRETACEOUS AND EOCENE BEDS.

AREA.

Sheets 49 S. and 50 S.E. of the Geological Survey Map include the towns of Aldborough (or Aldeburgh), Debenham, Framlingham, Orford, Saxmundham, and Woodbridge, and the important villages of Earl Soham, Grundisburgh, Leiston, and Wickham Market. The district is about 220 square miles in area, bounded by the North Sea on the East, but elsewhere by land.

RIVERS.

With the exception of a small brook rising near Tannington, on the northern edge of the map, and flowing north-westward to join the Waveney at Hoxne, all the district is drained by streams flowing to the S.E., and belonging either to the system of the Deben or to that of the Ore, except for the small streams N. of Aldborough. W. H. D.

The *Deben* rises N. of Debenham and flows S.E to near Rendlesham, receiving various tributaries on the way, from Helmingham, Kenton, and Debach. It then turns S.W., to its estuary at Woodbridge, being joined by a brook from Bredfield.

Its chief tributary is the *Finn*, which, rising N. of Witnesham, flows S. to Tuddenham (just beyond our district) and thence eastward to the estuary of the Deben, receiving in its course the brook that rises at Otley and runs past Grundisburgh and the Bealings.

The *Ore* (anciently Frome), rises near Saxtead, flows S.E. to below Marlesford and then eastward to the coast at Aldborough, after receiving the Alde, and at Snape the brook that, rising above Kelsale, flows S. through Saxmundham. At Aldborough the river, debarred by the shingle from joining the sea, turns S.W. along the coast, escaping at Hollesley, four miles below Orford and 10 from Aldborough, after receiving the Butley River.

The tributary *Alde*, rising near Dennington, flows S.S.E. into the Ore east of Little Glembam. Below this junction the names Ore and Alde are used indifferently on the old Ordnance Map.

The *Butley River*, rising south-eastward of Rendlesham, flows E. to Chillesford, and then S. to the estuary of the Ore, beyond our district.

The *Hundred River* is a small stream that rises near Knoddishall, and flows eastward to the sea N. of Aldborough.

The *Minsmere*, which drains the N.W. corner of the district enters the sea N. of Sizewell.

According to the report of the Select Committee of the House of Lords on Conservancy Boards (Fol. 1877) the Alde (= Ore) is $30\frac{3}{4}$ miles long, with a drainage area of 127 square miles, the Deben, 32 miles long, with a drainage area of 159 square miles, and the Minsmere is $12\frac{1}{2}$ miles long, draining 25 square miles.

W. W. and W. H. D.

The following extract from Reyce's Breviary of Suffolk in the Harleian MSS. (No. 3873) shows both the ancient nomenclature and the change that has taken place in the position of the mouth of the Ore :—

"Another *special* River, called of old time *Fromus*, ... [Ore] beginning at Tannington and Framlingham, descendeth to Marlesford, and so of the south-east of Farnham, entertaineth another River, called the Gleme [Alde], which cometh from Rendlesham [Rendham] and both the Glemhams, thus passing forth to Snape Bridge, it embraceth another River coming from Carlton by Saxmundham, and so continuing his course by Iken receiveth a third small brook, with all which accompanied, it fetcheth a great compass towards Aldeburgh and Sudbourne, and at length dedicates itself into the broad sea at Orford."

Physical Features.

The general form of the surface is that of a plain sloping eastwards, from 231 feet above the sea at Ash Bocking and about the same above Kenton to 36 at Sizewell Cliff and 78 at Sudbourn Black Walks, with a sharp descent to the sea-level at the estuaries.

The distinction between High and Low Suffolk is not a purely conventional line drawn across this slope, but rather a geological one separating the part where clay predominates from that which consists chiefly of light land. As most of the clay-surface is above the 100 feet contour-line and most of the sand below that level, the names bear with them a certain amount of geological correctness. Ignoring the valleys, which are merely gashes in the general plain, this boundary-line is approximately that of the railway; the principal towns and the heads of the estuaries being on it, through the operations of purely geological causes, the junction affording not only a starting point for denudation, but a

INTRODUCTION.

fertile mixed soil, an abundant water-supply close to the surface, and building-materials. This connexion obtains in fact over a large part of the Eastern Counties, most of the principal towns being on sites of peculiar (and favourable) geological structure or at the heads of estuaries.

GEOLOGICAL FORMATIONS.

The beds that occur at or near the surface in this district are as follows:—

Recent	{ Shingle. { Alluvium.
Post Glacial	- River-Gravel.
Glacial	{ Boulder Clay. { Gravel, Sand, Loam, and Boulder Clay. { Brickearth and Boulder Clay.
Pliocene	{ Chillesford Beds (Clay and Sand). { Red Crag. { Coralline and White Crag.
Lower Eocene	{ London Clay. { Reading Beds.
Cretaceous	- Chalk.

The Reading Beds have been proved by wells to occur, as elsewhere, between the London Clay and the Chalk. They are not, however, to be seen at the surface, owing to the covering of newer beds.

CHALK.

The Chalk is exposed in very few places, in only one of which does it exhibit its ordinary appearance of soft, white, earthy carbonate of lime. The other exposures (mostly natural and therefore weathered) present a hard, clayey, homogeneous mass, as if great pressure had been applied, destroying the granules of the original substance by welding or puddling them together.

In the N.W. part of the district the surface of the Chalk approximates in slope to the general contour of the country, lying but little below the bottom of the principal valleys, and at one point rising somewhat above the water-level. To the S. and E. it passes below the sea-level, the Tertiaries and Drift overlying it.

The inclined plane of the base of the Tertiaries, as fixed in the well-sections at Melton, Orford (Lantern Marshes), and Saxmundham (see Appendix), is found by protraction to pass but a foot or so above the surface of the Chalk at Framlingham, Easton, and Wickham Market. West of these places the Chalk has been denuded, forming a floor, on which rest the Crag and the Drift.

The puddled Chalk, above alluded to, is exposed in the small valley half a mile west of Framsden Meeting House, in a small pond, and is occasionally dug into, in draining, in the Deben alluvium between Winston and Ashfield

It is seen in drains and ponds at intervals from Earl Soham down to opposite Manor Farm on the east of the brook, and it is said to have been worked in the old brickyard to the west of the village. North of the cottages at Kings Hill, S. of Earl Soham, in an old pit on the western side of the road, the Chalk occurs in its normal condition, 10 to 15 feet above the level of the adjacent brook. In the small stream between Brandeston and Earl Soham, the puddled Chalk is exposed in several places, the highest being due west of Hill Farm, where it is about 25 feet above the level of the main brook. The Chalk is also found in draining the marshes below Monewden Hall.

A report that the Chalk comes near the surface at Broadwater, below Framlingham, seems to be based on the presence of a bed of chalk boulders rather than of the Chalk in place: it is, however, probably not more than 20 feet from the Alluvium anywhere between Framlingham and Marlesford.

Patches of reconstructed chalk also occur half a mile N.E. of Brandeston Church, and in the brook below Baddingham White House 2½ m. N.E. by N. of Framlingham; but it is doubtful if in either case the Chalk itself is near the surface.

READING BEDS.

A mass of mottled red and greenish loam on the bank of the Deben, between Melton and Woodbridge, appears to have been placed there to form the bank. As no such bed is seen in place* the material was probably dredged from the river-bottom. The blue and brown clays belonging to this series, brought up in boring a well at Saxmundham, were indistinguishable in character from parts of the London Clay. The thickness of the series may be taken at from 40 to 50 feet. The boundary probably trends from Otley eastwards by Charlsfield and Pettistree to the Deben, and then northwards, by Campsey Ash, Great Glemham and Rendham.

LONDON CLAY.

Of this deposit little more can be said than of the last series, though it is occasionally exposed. It is normally a stiff blue clay, weathering to chocolate brown, and its surface is generally rendered less tenacious by the admixture of sand washed down from overlying beds. It attains a thickness of 170 feet about Orford, but in our district its upper part has invariably been removed by denudation in the Miocene and subsequent periods. It is exposed at intervals in the Finn valley below Witnesham, in the next valley below Burgh, in the Deben below Ufford, and on the W. bank of the Butley River. Its boundary-line probably runs by the south of Otley, Rendlesham, Snape, and Leiston.

Sandy clay is seen on the road E. of Grundisburgh Hall. In the brickyard E. of Woodbridge the London Clay is bedded, and the sand dug from beneath it contains sharks' teeth. A quarter of a mile to the S.W., at a pond on the north of the railway, the clay was proved by boring to be 6 feet thick, over sand. In the brickyard on the railway N. of Wilford Bridge (Melton) the bedding of the London Clay (which is sandy in the lower part of the section) dips slightly to the S.W.

A quarter of a mile S.W. of Wilford Bridge the London Clay contains septaria and dips at a low angle to the S.E.

---

* From what seems to be the base of the London Clay having been found near the surface at Woodbridge, I was led to think that this mass of variously coloured mottled clay might be in place, brought up by a slight rise. Its occurrence at the edge of the marsh, by the water, as a detached mass, favours Mr. Dalton's view. See also the well section at the end of p. 55.   W. W.

CHAPTER II.—CORALLINE CRAG.

GENERAL DESCRIPTION.

The Coralline, White or Bryozoan Crag is the lowest division of the Pliocene formation in the British Isles, and is only found in the county of Suffolk. It is correlated by M. VAN DEN BROECK with the Middle Sands of Antwerp, and placed in the Diestian or Lower Pliocene, though more recent than the Sables de Diest.* It consists of beds, mostly light in colour, of shelly sand (locally called Crag) with thin bands of crystalline limestone, overlain by soft yellow rock largely composed of Bryozoa or Corallines. A third division is constituted by PROF. PRESTWICH and others, of the upper 6 feet of the Rock-bed, consisting of shell-detritus and sand; but no clear line of separation is visible, and the difference appears due only to frost-action.

The older buildings of the district frequently contain blocks of the Rock-bed, and the abandoned quarries often present a vertical face, weathered grey and sprinkled with moss and lichen, giving the appearance of the more compact Secondary rocks. The thickness of the series is not easy to ascertain but may be taken at 50 or 60 feet.

The nodule-bed at the base consists of nodules of phosphate of lime, rolled mammalian bones impregnated with the same mineral, and large rounded and subangular stones of various origin, with a few shells. Amongst the stones are generally to be found fossils derived from the London Clay and from other formations, sometimes phosphatized. The mammalian bones are supposed to be derived from some deposits, now destroyed, of an age between that of the London Clay and that of the Coralline Crag. The occurrence, in a similar bed at the base of the Red Crag, of pebbles of rock containing fossils of Diestian species, confirms this hypothesis. The knowledge of this base-bed of the Coralline Crag has however been derived from one section, closed many years ago, at Sutton, out of our district. (See Memoir on 48 N.)

With regard to the other divisions made by PRESTWICH (see p. 8), although their validity has been disputed, yet as they are partly lithological they may perhaps be retained, without attaching too great importance to them. W. H. D.

FOSSILS.

Although occurring over so small an area, and with an outcrop of only a few square miles, the Coralline Crag has yielded a

---

* Esquisse géologique et paléontologique des Dêpots pliocénes des Environs d'Anvers. Fasc. ii. *Ann. Soc. Mal. Belg.*, t. ix. (1878.) Pp. 130–136 of separate reprint.

larger number of species than all the rest of the Crag together, including even the many derived species of the Red Crag. It is, perhaps, the richest hunting-ground for fossils in the kingdom, not only from the number of species, but also from the abundance of specimens (in so many cases) and from their frequent perfection.

As with the Red Crag, authorities differ concerning the total number of species, PROF. PRESTWICH making this 559, in 1871,[*] with the proportion of living and extinct species shown in the table below, whilst in 1872[†] MR. BELL makes the number 707, to which he adds 38 in an Appendix to separately printed copies of his paper. These differences are owing, like those already noticed in the account of the fossils of the Red Crag,[‡] to the difference between naturalists as to what are specific or varietal characters.

The above-mentioned authors have given lists of these fossils, as also has Mr. WOOD, for the Mollusca, in his great work thereon. It is needless here to specify names, or even to note the more common genera; enough to say that from the tables given, and from the opinions expressed by various authors (see below, under Literature), the rich fauna of the Coralline Crag points to a milder clime than our own, and to conditions in which marine life was abundant.

PROPORTION OF LIVING AND EXTINCT SPECIES. (PRESTWICH.)

	Extinct.	Living.	Total.
Mollusca	52	265	317
Foraminifera	47	53	100
Bryozoa	65	30	95
Entomostraca, Corals, Cirripedia, and Echini.	34	13	47
	198	361	559

Of the above 265 species of living Mollusca the same author gives the following range, on the authority of Dr. GWYN JEFFREYS. In this case, and in the table from Mr. WOOD's work, I have added the percentage figures, so as to make comparison more easy:—

Mediterranean	-	-	200 species,	or over	75·4 per cent.
British	-	-	185 ,,	about	69·8 ,,
West European	-	-	171 ,,	over	64·5 ,,
Scandinavian	-	-	135 ,,	,,	50·9 ,,
Mid Atlantic	-	-	99 ,,	,,	37·3 ,,
Deep Atlantic	-	-	92 ,,	,,	34·7 ,,
Arctic	-	-	34 ,,	,,	12·8 ,,
Various others	-	-	12 ,,	,,	4·5 ,,

---

[*] *Quart. Journ. Geol. Soc.*, vol. xxvii., p. 134.
[†] *Proc. Geol. Assoc.* vol. ii., No. 5, p. 187.
[‡] The Geology of the Country around Ipswich, etc., p. 31 (1885).

Of the 391 species of Mollusca which he recognised in 1874 Mr. WOOD gives the following range:—*

British and Mediterranean	- 154 or over 39·3	per cent.
Mediterranean and not British	- 51 ,, about 13	,,
British and not Mediterranean	- 20 ,, over 5·1	,,
Not British or Mediterranean	- 24 ,, ,, 6·1	,,
Total living	- 849	63·6
Not living	- 142 about	36·3

These figures would be slightly altered by later additions in Mr. Wood's second and third Supplements (1879, 1882).

LITERATURE.

In the Memoir on the district to the south (48 N.) an account is given of the various works that refer to Red Crag geology, and, as many of these treat of the Crag generally, or as a whole, there is no need here to repeat the notice of their contents: we may confine ourselves to papers that specially deal with the Coralline Crag, and which are not noticed, or only partly noticed, in that Memoir, referring the reader to it for most of the general Crag papers, beginning with those of Mr. CHARLESWORTH, in 1835, the first in which the Crag was divided.†

Mr. R. FITCH, in a short note printed in 1835,‡ defended Charlesworth's term Coralline Crag, though on the mistaken ground that the abundant fossils are corals.

In 1853 PROF. E. FORBES and Mr. S. HANLEY, in the Introduction to their work on British Mollusca,§ remarked that the source of our molluscan fauna is that of the Coralline Crag, in which are to be found many of the ancestors of our living shell-fish, mostly forms of southern type. Some of these have lived on to our time, but most, struggling with the advent of less favourable conditions during the deposition of the Red Crag, were banished from our seas when glacial conditions set in, and did not return until the restoration of warmer times.

In 1854 Mr. S. V. WOOD described some tubular cavities in the Coralline Crag near Sudbourn and Orford.‖ These differ from the ordinary funnel-shaped pipes, which also occur (though of small size), are generally of 18 or 20 inches diameter throughout, and perpendicular, or nearly so. The walls of the most perfect were smooth, and masses of *Fascicularia* and *Theonoa* were cut through as if by a boring-tool. He thought that these chimney-pipes could not have been formed by the downward action of acidulated water (as the funnel-pipes have been); but were probably due to the upward issue of acidulated gas, while the Crag was beneath the sea. He alludes to some cavities near the bottom of the Coralline Crag at Ramsholt, of which I believe there is now no trace.

Dr. J. G. JEFFREYS, in his work on British Shells, concludes that nearly 60 per cent. of the marine shells of the Coralline Crag are of species now living in British seas, and that this formation is "the starting-point, and as it were the cradle of our molluscan race."¶

---

* Supplement to the Crag Mollusca, Part ii., p. 219.
† The others are WOODWARD, 1835 and 1836; CHARLESWORTH, 1836 and 1837; DESNOYERS, 1837; LYELL, 1839; WOOD, 1848; PRESTWICH, 1849; CLARKE, 1851; WOOD, 1859; CARPENTER, 1865; LANKESTER, 1865 and 1867; GODWIN-AUSTEN, 1866; JECKS, 1870; BELL, 1871 and 1872; LYELL, 1873.
‡ *Phil. Mag.*, ser. 3, vol. vii., pp. 463, 464.
§ A History of British Mollusca and their Shells, vol. i., p. xxxv.
‖ *Phil. Mag.*, ser. 4, vol. vii., pp. 320–326, pl. v.
¶ British Conchology, vol. i., pp. lxxxix–xcii. 8º. *Lond.* 1862.

In 1866, in a paper on the Red Crag,† Mr. S. V. Wood says, that in the Coralline Crag there "are the remains of 27 genera that are extinct in the British seas . . . . From this it is fair to infer that this Crag belonged to a period long antecedent to the deposition of the Red."*

In 1868 Mr. Godwin-Austen remarked of the Bryozoan Crag, as he calls it, that " it is a good division, because it is an indication of a definite range of depths, where the sea-bed was not within reach of surface-disturbance, yet where the drifting power was considerable, and having its own proper fauna. . . . . . Assigning to these beds depths of 40 fathoms, a difference of 300 feet is the least that can be assumed as that of their original, compared with their present conditions."†

Prof. Prestwich's paper on the Coralline Crag appeared in 1871.‡ In it the author remarks that the area of outcrop is only about 8 square miles (in which he probably includes parts where there is a covering of Red Crag), and that, though the Coralline Crag may have extended from Aldborough to Tattingstone, yet, if so, it has been removed by denudation, except in the low range of hills from Gedgrave northward to Orford, Sudbourn and Iken, and the outliers of Aldborough (this may be not an outlier but only separated by marsh), of Sutton and of Tattingstone. The surface of the London Clay, moreover, beneath the Coralline Crag is uneven.

He divides the formation as follows:—

		Feet.
Upper Division, 36 feet.	*h.* Sand and comminuted shells	6
	g. Soft false-bedded stone, made up of broken-up shells and remains of Polyzoa	30
Lower Division, 47 feet.	*f.* Sand, with many small shells and layers of broken-up shells	5
	e. Sands, with many Polyzoa, often in the position of growth, some small shells and *Echini*	12
	d. Broken-up shells and large whole shells, with layers of limestone in the upper part	15
	c. Marly beds, with many well-preserved shells, often in the position in which they lived	10
	b. Broken-up shells, Cetacean remains and Polyzoa	4
	a. Phosphatic nodules and mammalian remains	1

Note.—The thickness of *e* may be 2 feet too much, that of *d* 5 feet, and therefore that of the lower division 7 feet. (See later paper in same vol., p. 496.)

Details of sections and lists of fossils are given, and a lengthy review of the fauna, with its relations to existing faunas, from which it is concluded that the differences in the proportions of recent to extinct species in the different classes is so great that the results are difficult to reconcile.

The history of the formation is thus traced out: Between the period of the London Clay and that of the Coralline Crag our Eastern Counties seem to have been dry land, though to the south and east there was sea, which gradually encroached westward, over our area: a movement which was accompanied by a rise of the land to the east, and perhaps to the south. With this movement the climate got colder, as evidenced by boulders (probably ice-carried) in bed *a*. Subsidence went on, finer materials (*b*) were deposited, and then *c* and *d* were formed in comparatively deep tranquil water, being succeeded by the Polyzoan bed (*e*) showing the greatest depth of the period (from 500 to 1,000 feet). Then came a slight shallowing, with the deposition of *f*, and further elevation, exposing the sea-bed to the action of tides and currents, led to the heaping up of the upper division. The continuance of the movement of elevation raised the Coralline Crag above the sea and exposed it to denuding actions, which broke it up into islands and reefs.

A general lowering of temperature, or, more probably, the setting in of fresh currents from the north, from the continued subsidence in that direction, led

---

* *Quart. Journ. Geol. Soc.*, vol. xxii., p. 541.
† *Geol. Mag.*, vol. v., pp. 475, 476.
‡ *Quart. Journ. Geol. Soc.*, vol. xxvii., pp. 115–146, pl. vi.

to the introduction of northern forms of life and to the gradual extinction of southern ones.

In 1872 Messrs. S. V. WOOD, JUN., and F. W. HARMER* regard the Coralline Crag as not having a thickness of over 60 feet. They doubt the constancy or determinability of the horizons into which Prestwich divides the Lower Division; for, so far from their being characterised by groups of fossils, Mr. Wood's long researches have been mainly confined to one pit at Sutton, with a vertical range of only a few feet, and from which he has got specimens of nearly all the species of the formation. Moreover, so inconstant is the Molluscan fauna, that many species once found at a place may not be noticed there again for years. They object also to the depth of deposition assigned to the Polyzoan bed (e), as it would have carried the Crag sea over all East Anglia, and beyond, and it is unlikely that all traces of such a sea should have been removed. Again, nothing among the Mollusca points to a greater depth of water than 40 fathoms.

Two years later, in 1874, Mr. S. V. WOOD† remarked that the Molluscan fauna of the Coralline Crag has Mediterranean affinities, and that the conditions of temperature of the period, as inferred from the Mollusca, seem to have been nearer to those of the seas of Southern Europe and of the Azores than to those of British seas. The most abundant species are southern ones, but there are many species of arctic or boreal character.

Mr. P. F. KENDALL, in 1883,‡ drew attention to the fact that whilst the shells in a pit near Aldeburgh were, with one exception, of the kinds determined by Dr. Sorby to have their carbonate of lime in the calcite-form, the many casts were, without exception, of the kinds in which the carbonate of lime is in the form of aragonite. The former, of course, is the more stable, and the latter the more soluble. The calcite-shells of the Crag, moreover, have been almost wholly free from the attacks of boring animals. The consolidation of the Coralline Crag, and the dissolution of its aragonite-shells, are concluded to have taken place before the deposition of the Red Crag, a fragment, with casts of shells, from the older bed having been found in the newer one.

In 1885 Mr. W. H. DALTON noticed§ the manner of infilling of some casts of *Voluta Lamberti*, but his remarks have been reproduced further on (p. 11).

In the Memoir on the district to the south (the northern part of Sheet 48), the outliers of Tattingstone, Sutton, &c., are described.   W. W.

DETAILS.

A part of the principal mass of the Coralline Crag, extending into Sheet 48, N.E., has been purposely left undescribed in the Memoir on that district, in order that the whole mass might be dealt with in the following pages.

In Boyton Marshes, west of the mouth of the Butley River, the phosphate-nodule beds of the Coralline and Red Crags are in contact,|| the upper part of the Coralline Crag having been removed by erosion. A little further north, a shallow pit was worked in 1871, near Bush Covert. The exact spot is no longer identifiable, all trace of the phosphate-workings being obliterated by agriculture, but PROF. PRESTWICH writes that "from the abundance of *Cardita senilis*, *Astarte Omalii* and *Cyprina islandica*, the occurrence, although rare, of *Mytilus hesperianus*, *Pecten maximus*, and *Isocardia cor*, and the absence of the ordinary shells of the Red Crag, with the exception of a few specimens of *Trophon antiquus*, near the surface, I should feel disposed to consider this a disturbed portion of the Coralline Crag, and to refer the 2-foot coprolite-bed below it to this formation."¶

As but two feet in thickness, at most, of the Coralline Crag remains undisturbed, it is, of course, impossible to indicate its presence on the map, except

---

* Supplement to the Crag Mollusca (by S. V. Wood), Part 1. *Palæontograph. Soc.* Pp. ii-iv.
† Supplement to the Crag Mollusca. Part ii., pp. 192-196.
‡ *Geol. Mag.*, dec. ii., vol. x., pp. 497-499.
§ *Quart. Journ. Geol. Soc.*, vol. xli., *Proc.*, p. 2.
|| See The Geology of the Country around Ipswich, etc., p. 28.
¶ *Quart. Journ. Geol. Soc.*, vol. xxvii., p. 125.

by the sign k^1 engraved on the boundary-line between the London Clay and the Red Crag.

The shelly sands of the Coralline Crag are well seen in two pits at Gedgrave, one about 200 yards south-east of Ferry Barn, the other at Low Farm, both showing sand with layers of comminuted shells and irregular bands of shell-limestone, probably belonging to Prof. Prestwich's division d.

The cattle-yards at Gedgrave High House are excavated in the rock-bed g., but the vertical sides, overgrown with lichen, liverwort and mosses, do not offer favourable opportunities for examination.

The Gomer pit, which in 1863 afforded DR. S. P. WOODWARD so rich a series of Mollusca, was near the edge of Sheet 48, N.E., a quarter of a mile east of the Butley River, and in division f.

The pit near the Keeper's Lodge, Broomhill, west of Orford, is thus described by PROF. PRESTWICH :—*

Surface and Drift soil, 3 feet.
Yellow sand, full of detached Bryozoa, chiefly *Fascicularia* and *Alveolaria*, a few shells (e), 7 or 8 feet.
Sandy beds of comminuted shells, intercalated, in which are layers of large shells, well preserved and often double (in the lower part) a few Bryozoa, and thin bands of tabular limestone (d), 15 feet.

The rock-bed (g) has been quarried, through overlying beds of red sand, near Orford Castle, and (without covering) near Roydon Hall Farm, where it extends to the sea-level, but rises again northward, the shelly sands f being seen at the edge of Sudbourn Marshes, east of Ox House, and the rock-bed being quarried close under the Chillesford Clay feature, a little to the south of Ox House.

In Sudbourn Park there are some shallow pits, almost wholly weathered down, about half a mile south-west of the church and close to the road into Orford. The rabbits burrowing into the decomposed rock throw out large numbers of *Fascicularia*, *Alveolaria*, &c., of large size and in perfect preservation.

There is a fine pit in the angle of the roads a third of a mile W.S.W. of Sudbourn Church, and another, abounding in Echinoderm remains, in the Park, about a quarter of a mile N.N.E. of the Hall. Near the stables the division d is seen in a large shallow pit : the shelly sands contain many large *Cyprinæ*, and other lamellibranchs with united valves, and the surfaces of the thin irregular bands of limestone are covered with delicate *Polyzoa*, indicating probably the contemporaneous deposition and solidification of the stone.

The rock-bed is, or has been, quarried at several points along the base of the Red Crag to the east-north-east of Sudbourn Church, but in no case are the exposures of the lower series of interest, except as determining the limits of the upper. At the cross roads, four-fifths of a mile to N. 28° E. of the church, is a large pit showing PROF. PRESTWICH's division h, the upper 6 feet being loose shelly sand with few *Polyzoa*; but this appears to be merely a disintegrated condition of the rock-bed. Half a mile east-north-east of this are sands, probably of division e, crowded with *Polyzoa*.

These beds, or those of division f, skirt the marshes to Webber's Whin, rising gradually northward. They are seen at Calton Farm, and again half a mile to the west, and then descend northwards. The rock-bed is fairly exposed at Iken brickfield, about half-a-mile W.N.W. of Calton Farm (see p. 21); and the beds f near Redland's Covert. The latter here consist of firm stony Crag, a mass of shells and casts, false-bedded at top and bottom, but not in the intermediate beds. Among the shells *Mytili* largely predominate, forming here and there the entire mass of considerable slabs. Near the top are concretionary masses, and a few Polyzoa. Bands of tufa occupy crevices and bedding-planes of open texture.

On the north of the Ore, opposite to Stanny Point, similar beds form a low cliff of loose brashy rock, and a shallow pit, a quarter of a mile northward, is in like material.

From a small pit in the rock-bed, about 100 yards N. of Aldborough Hall, were obtained many casts of *Voluta Lamberti*, the inner whorls of which were

---

* *Quart. Journ. Geol. Soc.*, vol. xxvii., pp. 122, 123.

cut off by planes representing the surface of the fine calcareous mud within the dead shell as it lay at the bottom; the outer whorl only receiving its full complement of sediment, which rose to less and less height in each successive whorl. (Figs. 1, 2.)

The planes really are somewhat curved, rising slightly at the margin through the capillary attraction of the fine mud to the side of the shell, then there is a ring of depression, and then an upward arching of the interior. These curves point to the presence of an elastic cushion of gases (evolved by the decomposition of the animal) acted upon by varying pressure (probably tidal) communicated through the mud from the mouth of the shell. Unfortunately the curvature of the surfaces has been indistinctly rendered in the cut.

Figs. 1 and 2.

Casts of Voluta Lamberti, *showing the Filling in of the Whorls of the Shells.*

Drawn by Mr. J. G. Goodchild, from specimens in the collection of Mr. H. Stopes.

Near the Red House there are pits on both sides of the high road, in which, as elsewhere, the formation of pipes, filled with the reddish-brown sandy earth resulting from the decalcification of the rock may be studied with advantage. The effect of these pipes on overlying beds will be referred to later on. (See p. 26.)

This is the most northerly point at which the Coralline Crag is seen, but Mr. C. P. Ogilvie of Sizewell tells me that it forms dangerous sunken rocks off Thorpe and Sizewell.

CHAPTER III.—RED CRAG.

General Description.

The Red Crag consists normally of very shelly sand, of a blue colour, from the contained protoxide of iron; but in this form it is only met with in sinking wells, in places where the shells and the colouring-matter are alike protected from the action of the carbonic acid and oxygen contained in percolating water. Nearer the surface the blue is changed by oxidation of the iron into red, and the shells are rendered very friable. But the most usual form of this deposit, at, and often for 20 feet below, the surface, is sand of a deep orange-brown colour, with bands and streaks of clayey peroxide of iron, sometimes passing into hard, blackish-brown limonitic concretions, solid or hollow, and, in the latter case, empty or partially filled with angular fragments of ochre. The concretionary structure is sometimes developed on a very large scale, affecting the beds through three or four feet of vertical thickness, and 30 or 40 feet in horizontal extent, as in Fig. 4, p. 13. The bands and streaks of varying intensity of colour, whether, as here, in concentric circles, or, as is generally the case, horizontal, are clearly of posterior date to the deposition of the beds, and are due to the re-arrangement of the impalpable particles of oxide of iron by percolating water. The unaltered shelly beds are often very obliquely current-bedded, and a mass of this nature is sometimes seen rising as a boss in the middle of the altered sand, the apparent bedding of which is horizontal, as in Fig. 6, p. 23.

These bosses of unaltered shelly Crag sometimes graduate into, at other times are sharply separated from, the decalcified part. In Fig. 3 the concentration of ochreous material at the limit of percolation is instructively shown.

Fig. 3.

Section in a Pit, 3 Furlongs East of Great Bealings Church.
Messrs. Wood and Harmer. Quart. Journ. Geol. Soc., vol. xxxiii., p. 76.

Scale 10 feet to the inch.

b. Red stratified sands, being *a* altered and restratified.
b'. Band of dark, partly hardened, ferruginous loam.
a. Red Crag, unaltered and full of shells.

FIG. 4.

Section in a Pit on the Common about half a mile West of Butley Abbey.—PRESTWICH.*

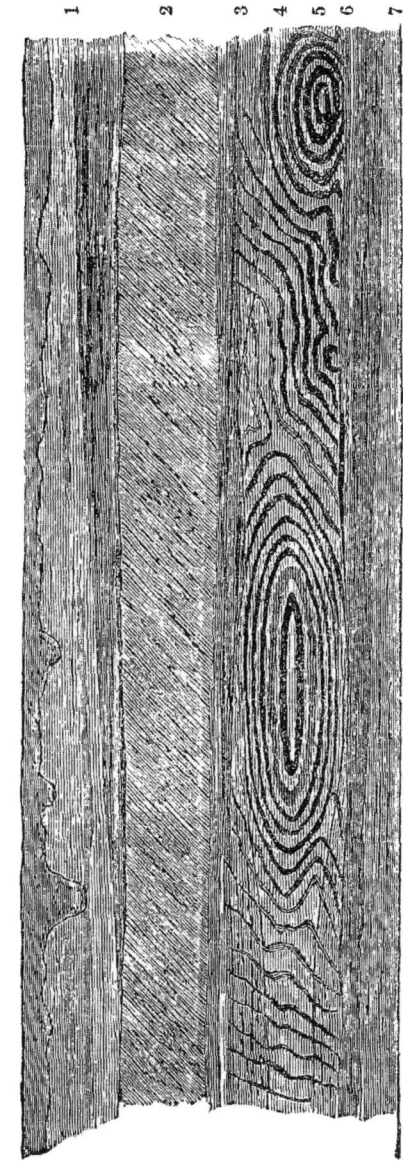

32 feet long and 10 feet high.

1. Light-coloured Crag.
2. Very ochreous Crag.
3. Ferruginous Crag and iron-sandstone.
4. Shell-bed.
5. Light-coloured horizontal Crag, with ferruginous concentric bands.
6. Shell-bed.
7. Ferruginous sand and iron-sandstone.

---

* Quart. Journ. Geol. Soc., vol. xxvii, p. 329. For the use of this and other woodcuts (Figs. 3, 5, 6) we are indebted to the courtesy of the Council of the Geological Society.

In many cases parts of the limonite have been formed before the decalcification was complete, so that casts and impressions of some of the shells remain to testify their former abundance.

The Red Crag has been subdivided as follows by Mr. S. V. Wood, jun., from structural and palæontological characters:— 3. Butley Crag; 2. Orwell-Deben Crag; 1. Walton Crag (the oldest).—The second of these contains only worn and *remanié* shells of the Walton Crag, and is without any trace of those that characterize the Butley Crag. He further thinks that there is reason to suppose that the region between the Orwell and the Deben was elevated above the sea-level before the formation of the Butley Crag, which was deposited on the shores of the newly-raised land. The boundary, however, of the divisions, if discoverable, being so only by laborious examination of the fossils obtained from sections separated by considerable stretches of decalcified sand, it has been found necessary, without impugning the correctness of Mr. Wood's deductions, to treat the Red Crag as a whole. The divisions, at most, are of less importance than many of the breaks in the older rocks which have not been honoured by more than a passing notice. Contemporaneous oscillations are also recorded in the Belgian Pliocene deposits.*

Details.

Valley of the Finn.

The most westerly point in this district at which the Red Crag is exposed is at Witnesham Street near the head of the Finn Valley. Here, in three small pits on the N. side of the high road, reddish current-bedded gravel and sand of Glacial age overlie a whitish sand presenting the appearance, not easily described, of decalcified Crag, *minus* its colouring matter. The usual ironshot sand is dug in a fourth pit about 200 yards S. of the public-house on the by-road to Tuddenham.

In the valley of the unnamed tributary of the Finn rising near Otley, several exposures of Crag occur.

From the high road at Great Bealings to the farms 700 yards W.N.W. of the church, shelly Crag is frequently seen in casual sections.

A large pit at the farm a third of a mile E. of Grundisburgh Hall is in the decalcified sand with a capping of Glacial gravel.

The shelly Crag is seen in an old pit a quarter of a mile E. of Grundisburgh Meeting House, but north and east of this, for several miles, only decalcified ferruginous sands are visible, though under the wide plateau of Boulder Clay, patches of unaltered Crag may exist.

In the side valley half a mile W. of Grundisburgh Church the following section is exposed in a sand-pit:—

Glacial.—Horizontally-bedded white sand, 18 feet.
Red Crag.—Obliquely-bedded ferruginous sand, 7 feet.

In a large pit just N. of Clopton Church, the decalcified sand, strongly current-bedded and with limonite-bands, is affected by many small faults, possibly due to settlement from unequal dissolution of shells. No trace of these, nor of phosphatic nodules, is seen in this or in the adjacent pits west of the high road and on the opposite bank of the brook descending from Clopton Common.

---

* Van den Broeck. Esquisse géologique . . . des Depôts pliocénes . . . d'Anvers, *Ann. Soc. Mal. Belg.*, t. ix. 1876, 1878.

The ferruginous sand is seen in several pits about Hasketon and Great Bealings, viz. :—

⅛ mile E. of Hasketon Hall	700 yards N.E. by N. of Hasketon Church.
⅕ „ S.E. „ „	- 600 yards E. by N. of Thorp Hall.
⅓ „ E. by S. „ „	- 200 yards N. by E. of Hasketon Parsonage.

The last of these shows sand and gravel upon current-bedded limonitic sand.

The road-cutting up from Thorp Hall is in decalcified sands, but half a mile east are pits, on either side of the road, showing shelly Crag, with a greater or less thickness of ferruginous sand over it, capped by sand and gravel. The hill-flank between Thorp Hall and Great Bealings affords several exposures of shelly Crag, as do the cuttings on the high road. The pit at the farmstead 200 yards N.E. of the church shows sand and gravel over shelly Crag.

Fig. 3 (p. 12) shows the structure of the Crag in a pit 750 yards eastward of Great Bealings Church, at the edge of the map. 200 yards N. of Bealings House a pit by a wood shows: Boulder Clay, over Yellow Sand, over Dark Red Sand (? Crag).

Valley of the Deben, up the Right Side.

At Woodbridge the decalcified ferruginous sands occur in a pit a quarter of a mile west of the Abbey, and shelly Crag has been found in well-sinking on Market Hill. At the back of the British Schools (400 yards N. by E. of the church) a fine pit gives the section below :—

Glacial Drift. Whitish sand and gravel alternating with red bands. abundant shell fragments; about 40 feet.
? Crags. Clayey red-mottled sand strongly current-bedded; 10 feet.

The occurrence of phosphatic nodules, in the rain-wash overlying the London Clay of the brickyard at the eastern end of Woodbridge, testifies to the proximity of the Crag, probably on the road just above.

In the hollow between Woodbridge and Melton Street shelly Crag is frequently exposed by rabbits, &c. It has been dug at the entry of the drive to Fern Villa.

On the western side of the grounds of Foxburrow Hall, west of Melton, a large overgrown pit shows :—

On the western side:
Slightly laminated Boulder Clay ⎫ Nearly vertical,
Coarse light-coloured laminated sand ⎭ dipping to west.

In the middle of the wide floor:
Decalcified ferruginous, over shelly, Crag.

On the north-eastern side:
Brown brickearth, slightly laminated (may be decalcified remains of Boulder Clay).
Chalky gravel and laminated sands, with tufaceous matter along beds of more porous nature.

The shelly Crag is seen in pits on the lane leading from the old turnpike-gate eastward to the brick-kiln. These show the usual decalcification in the upper part, and in one of them, blocks of consolidated shelly material, deeply stained with iron, are crossed by veins of white calcite, indicating more gradual deposition than the usual tufaceous deposit, and showing also the complete peroxidation of the iron and the consequent absence of any soluble ferruginous matter.

The pit 600 yards S.W. of Melton Church has been figured by Messrs. Wood and Harmer,[*] before the principles of decalcification and oxidation were recognized, as an example of a mass of Crag undermined and partially surrounded by Glacial beds. It is now known that the whole is of Crag age,

---

[*] Supplement to the Crag Mollusca, Part I., p. xxi., *Palæontograph. Soc.* 1872.

but the figure well illustrates the capricious manner in which the alteration often takes place. The altered part yields occasional casts of shells.

From Melton to Ufford the Crag does not come to the surface, but it is exposed, under gravel, in a small pit at the fork of the roads N. of Ufford Place.

Some red sand, probably Crag, is seen in the lane to Ufford Thicket.

An old pit a quarter of a mile S. of Byng Hall, in the side-valley, N.N.W. of Ufford, shows the following beds:—

Glacial Drift. { Boulder Clay.
{ Sand and Gravel, 4 feet.
Shelly Crag, 12 feet.

Of two pits north of Byng Hall the lower presents the following section, whilst the higher shows only the last two beds:—

Glacial Drift. { Boulder Clay.
{ Sand and gravel.
Decalcified Crag (red sand).

A pit a quarter of a mile E.S.E. of Java Lodge, southward of Pettistree, furnished in 1878 a rather complex section of Drift over Crag.

Eastern face of pit:

	Inches.
7. Grey-brown loamy sand	6-12
Line of gravel.	
6. Orange-coloured sand, passing S. into yellowish over white sand	42
5. Seam of blackish laminated clay.	

(This and the beds below it are broken by a small fault, trending N. 17 E. and shifting the southern beds down 12 inches: no break is noticeable in the beds above. A pocket of gravel above the black clay on the downthrow side only.)

4. Boulder Clay, with a seam of sandy brickearth in the middle - 30
3. Lenticular mass, to left side (N.) only, of coarse chalky gravel, over which the clay arches.
2. Orange-brown sand with limonitic stalactites and a few shells.
1. Yellowish-white and orange-coloured Crag, horizontally-bedded.

The right (S.) side of this section is hidden by a slip, and beyond it is the following:—

Gravelly Boulder Clay in position of (7):	
Orange-brown sand = (6)	48—42 in.
Blackish laminated clay = (5)	8 ,,
Boulder Clay = (4)	24—18 ,,
Orange-brown sand = (2)	4—10— 0 ,,
Boulder Clay	0— 2—12 ,,
Sand as (2)	12—14—16 ,,
Boulder Clay	2 ,,
Sand as (2).	

Between Ufford and Ash Abbey several exposures of shelly Crag occur on or near the railway, and the outcrop is traceable to the east of Pettistree, where the Glacial Drift descends into the bottom of the valley.

Red sand, probably decalcified Crag, occurs on both sides of Potford Brook, between Letheringham Lodge and Thorp Hall, and shelly Crag is exposed, under Boulder Clay, at the edge of the Alluvium east of the lodge, and also at Letheringham Old Hall, the preservation of the shells in these two cases being clearly due to the protective influence of the clay-covering.

Between Monewden, Winston, and Earl Soham the Crag has been proved to be absent by the laying bare of the Chalk floor, but this is due to denudation, as the Crag exists in full thickness ten miles to the north and is found at higher levels southwards.

Valley of the Deben, down the Left Side.

Traces of Crag were found in sinking the Artesian well in Easton Park (see p. 50), and west of the park a large pit showed the following section (1875):—

>Boulder Clay, on the northern side : a trace.
>Very sandy gravel, 15 feet.
>Soft ferruginous band.
>Fine soft grey micaceous current-bedded sand, 4 feet.

Mr. S. V. Wood, junior, has suggested that this last bed may be referable to the Crag.

A quarter of a mile west of Glevering Hall there is slightly ferruginous sand, with seams of soft very micaceous iron-sandstone, but without any trace of shells or of phosphate.

Similar sand occurs about a mile to the E.S.E., at the bend of the river below Gallows Hill.

From westward of Campsey Ash there is a continuous outcrop on the left bank of the river.

A pit in the cliff facing the Alluvium, a mile E.S.E. of Wickham Market, gives the following section:—

>Boulder Clay, 2 feet.
>Sand, 14 feet.
>Shelly Crag, 15 feet.

In the cliff facing the Decoy another pit showed 11 feet of sand over shelly Crag, 23 feet.

A large pit at the cross-roads above this shows only Crag, shelly below, limonitic above.

Shelly Crag is also seen in the railway-cutting near by, and in pits at Rendlesham High House and Naunton Hall. A pit about a quarter of a mile south of Bridge Farm, W. of Eyke, yields a considerable variety of shells. Over 30 feet of ferruginous sand is shown in a fine pit at the southern end of Wilford Bridge, and at a lower level the shelly Crag, which is also exposed at frequent intervals from thence to Sutton Haugh.

Valley of the Butley River.

One of three fine Crag pits between Butley and Tangham Folly to the S.W. has been figured above (Fig. 4). Much of the ground about Bush Covert has been worked over for coprolite, the layer of which was at one point 3 feet thick. It has already been stated that this phosphate-bed may belong partly or entirely to the Coralline Crag.

Several Crag pits lie between Carman's Wood (E. of the church) and the Water Mill, and between the latter and Staverton Park. Some of these have been described by previous writers, but from indefiniteness as to exact locality the sections cannot now be identified with certainty.

Dr. J. E. Taylor gave in 1871 a section at Butley as follows:—*

>Chillesford Clay.
>Upper Crag.
>Chillesford Clay.
>Crag.
>Lower Crag, false bedded.

This seems doubtful, no evidence of such division of the Chillesford Clay having been elsewhere noticed: possibly a slip of the face of the pit misled the observer. The probable position of this section is a quarter of a mile east of the Oyster Inn, almost directly above the pit which has been called the Butley Oyster pit, and which is thus described by Mr. A. Bell†:—"The section now presented by the excavation . . . . . has been cut into a

---

* *Geol. Mag.*, vol. viii., p. 314.
† *Geol. Mag.*, vol. viii., pp. 450, 451. (1871.)

gently rising slope for nearly 300 feet in length, by about 35 feet in its deepest part."

1. Drift sand, black, full of small particles of quartz, with a large number of rolled stones [Rainwash].
2. Red sand, passing into
3. Yellow sand, then into
   Red sand. A large mass of brownish clay (full of casts of the common mussel and *Trochus cinerarius*) at one part.
4. Vein of fine white sand, extending rather more than half-way across the pit.
   "These sands [2, 3, 4], etc. are about 18 to 20 feet thick, and are nearly, except where in contact with the Crag, *totally devoid* of organic remains. Finely comminuted shells occur at the lines of junction."
5. Red Crag full of shells.
6. Red Crag   ,,   ,,
7. Unfossiliferous sand, partly false-bedded at top, separating 5 and 6 [? both over and under 5, over 6 in the figure given].
8. Line of freshwater shells.
9. Red Crag with shells. "A series of layers of fine sand and shells, having a rapid dip, being the lowest deposit seen."
   "The cross section [nearly at right angles] gives traces of considerable erosion."

A list of 192 species of shells from Crag at Butley is given in this paper.

The Crag ceases to be shelly a quarter of a mile W.N.W. of the Oyster Inn; but is seen in its decalcified state up the valley to Orphans Piece and down the other (left) side to Wantesden Heath, between which and Chillesford there are several exposures of shelly Crag, one being in a hillock surrounded by alluvium. On both sides of the valley decalcification commences with the disappearance of the protecting sheet of Chillesford Clay.

At Chillesford, the pit, behind the church and that in the stackyard below the church, furnish an almost continuous section, PROF. PRESTWICH having proved by excavation the nature of the few feet of beds between the floor of the upper and the top of the lower pit. It must, however, be borne in mind that the latter pit is exposed to decalcification, from which the beds seen in the higher are exempt. The section is as follows :—*

Upper pit, by the Church. (16 ft.)
- Light-coloured Boulder Clay with a seam of broken shell-fragments at the base.
- Grey clay, with a few shells and fish, vertebræ, passing down into light-coloured clayey sand, with patches of perfect but friable shells.
- Yellow sands without shells.
- Part proved by digging. Yellow sands with few shells.

Lower pit, in the Stackyard. (22 ft.)
- Seams of ferruginous sands with a few seams of clay and ome shells.
- Seams of comminuted shells.
- Pebbly sand.
- Light-brown sand and iron-sand, with shells in greater variety and more perfect.
- Beds of comminuted shells with some entire.

At the brickyard, half a mile E. of the church, PROF. PRESTWICH noted in 1849 the following section :—

Yellow and grey laminated sandy clay, with indistinct casts and impressions of shells, 12 feet.
Yellow sand, with a few shells in the lower part, 5 feet.

In the well below this the shelly Red Crag was met with.†

---

* *Quart. Journ. Geol. Soc.*, vol. xxvii. p. 336.
† *Quart. Journ. Geol. Soc.*, vol. v., pp. 346, 347.

Prof. Prestwich mentions the occurrence of the vertebral column of a whale 31 feet long in the Chillesford Clay.*

South of the Decoy a few casual exposures of shelly Crag occur, and at Chillesford Lodge, and half a mile westward thereof, there are Crag pits. The Crag extends from these to half a mile W.N.W. of Sudbourn Church, but there are no permanent exposures in that space.

The Red Crag of the Orford outlier, between the Butley River and the Ore, is almost entirely decalcified, but a trace of shells is seen half a mile S. by W. of Sudbourn church.

---

* *Quart. Journ. Geol. Soc.*, vol. xxvii., pp. 337, 338 ; and Dr. Crisp, *Rep. Brit. Assoc.* for 1868, *Sections*, p. 61.

CHAPTER IV.—RED CRAG—*continued.*

DETAILS.

Valley of the Ore (or Alde), up the Right Side.

A fine section of ferruginous sands, resting upon the rock-bed of the Coralline Crag, is to be seen at Orford Castle.

A mass of white current-bedded sand caps the little hill east of Roydon Hall, and may possibly be referable to the Red Crag, which occurs below the shingle in the marshes east of the river. (See WELL-SECTIONS, Lantern Marshes, p. 53.) At Ox House there is no evidence of any bed separating the Coralline Crag and the Chillesford Clay, but the junction is not actually exposed, and there may be a foot or two feet of sand between them, representing the Red Crag.

On the eastern side of Sudbourn Common the Red Crag thickens out again, owing to the depression of the Coralline Crag surface. The two series are seen in junction in the following five pits:—

0·6, 0·8, 1·2, and 1·3 miles E. 20° N. from Sudbourn church, and 1·1 mile E. 35° N. from the church.

In the second of these the Red Crag, full of shells, is nearly white, whilst the Coralline or White Crag is deeply stained with iron-oxide, a testimony to the triviality of colour-names. A band of phosphatic nodules occurs at the junction, and the current-bedding of the Red Crag dips at 30° to W.

There seems to be some error in the designation of the locality given by PROF. PRESTWICH to the section shown in Fig. 5, as there is no trace of a pit at that point, and the Chillesford Clay there separates the Glacial sand from the Red Crag. If for $1\frac{1}{4}$ we read $\frac{3}{4}$, there is a pit which might have once shown such a section: an irregular junction of the Red and the Coralline Crags.

FIG. 5.

Section in a Pit by the Side of the Road $1\frac{1}{4}$ (? $\frac{3}{4}$) mile N.N.E. from Sudbourn Church.

(PRESTWICH. *Quart. Journ. Geol. Soc.*, vol. xxvii., p. 335.)

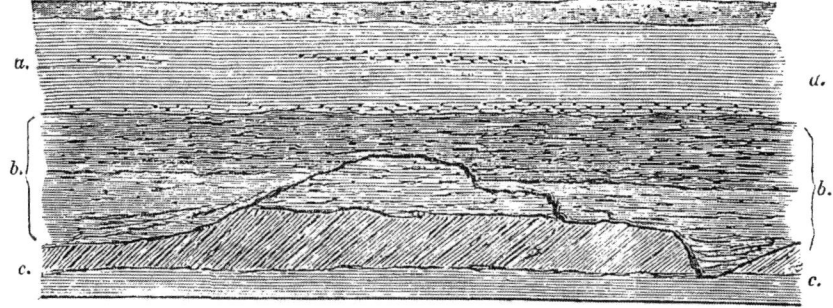

a. Light-coloured sands with fine gravel.
b. Ferruginous and yellow sands (Red Crag).
c. Coralline Crag.

From this point the Crag again becomes very thin, but appears to be continuous. It is seen at several points from Webber's Whin to the Iken Brickfield, where it is only 4 feet thick.

The REV. O. FISHER notes of a pit on Webber's Whin, now sloped down, that the shells in the sand under the Chillesford Clay were mostly double, and in the position of life.*

At the brickfield (1⅜ mile S.E. by E. of Iken church) the following section is recorded by PROF. PRESTWICH (1849) :—†

	FEET.
Flint gravel, on eroded surface	1–3
[Chillesford Clay]. Laminated grey clays and sands with indistinct impressions of shells [The upper 4 feet of the clay is partly resorted. W. H. D.]	10
[Red Crag]. Yellow sands, gravelly at the base	4
Light bright yellow calcareous Coralline Crag with its top surface slightly uneven and	over 30

The Red Crag ranges in this attenuated form along the hillside to near Redlands Covert. Here the lowering of the Coralline Crag surface re-introduces the shelly beds, which are seen in several pits in the side-valley descending from Tunstall. The upper part is, however, in most cases decalcified. The following are the principal exposures :

200 yards S.E. of keeper's house.
600 „ S. „ „
S.W. side of the road from Snape to Sudbourn, on both sides of the stream.
1¼ mile N.W. of Sudbourn church (E. of wood).
1½ „ W.N.W. „ „ (N. of road).
300 yards W. ⎫ of the last.
700 „ N.W. ⎭

The second of these is thus described by PROF. PRESTWICH ‡ :—

Chillesford Clay, with casts of shells ; 10 feet.
Light-coloured sand, passing down into shelly Crag.

Near Tunstall Meeting-house, a pit on the common exposes—

Peaty sand.
Light-coloured sand, with gravel at the base ; up to 6 feet.
Light and bright rusty-coloured, bedded and false-bedded sand, with fine loamy layers (Crag).

The ferruginous sand is also met with in the bottom of a large old pit among the houses.

The belt of decalcified Crag skirting the east of Iken Heath affords no good sections south of Iken Hall. A fine exposure is formed at the bend of the river at Iken Cliff, where the ferruginous sand has been regarded by Mr. S. V. WOOD, jun., as part of the Glacial Series.§ Near the cross-roads, three quarters of a mile W. by S. of Snape Bridge, shelly Crag is again seen.

Valley of the Ore.

Near the junction of the Ore and Alde the Glacial Drift descends to the level of the valley for a few hundred yards, but from Beversham Bridge to Marlesford there is, on the right bank, an uninterrupted narrow belt of Crag, shelly at the base, and decalcified above. The upper beds are well seen in pits at Blackstock Wood and west of Marlesford Bridge, beyond which no trace of the Crag Series is found on the western side of the Ore valley.

In the artesian well at Framlingham College no trace of the Crag was present, but seven eighths of a mile S. of the station, the railway-cutting on the left side of the valley touches ferruginous sandstone, in which a shell was found, and which may represent the Crag.

---

* *Quart. Journ. Geol. Soc.*, vol. xxii., p. 19.
† *Quart. Journ. Geol. Soc.*, vol. v., p. 347.
‡ *Quart. Journ. Geol. Soc.*, vol. xxvii., p. 338.
§ *Quart. Journ. Geol. Soc.*, vol. xxxvi., pl. xxi. (1880.)

A large pit, 170 yards N. of Parham Station, gave the following section (1876) :—

Drift.
- Boulder Clay (eastern side only).
- Light-coloured bedded sand with three-inch seams of grey clay : 10 feet.
- Gravel, with pebbles of ironstone : 6 inches to a foot.

Red Crag. Reddish-brown bedded sand, with much ironstone, one cast of a shell : 12 feet.

W.N.W. of Parham Hall the railway-cutting reached, at its northern end, deep reddish-brown sand, resting on compact ironstone, crowded in places with impressions and casts of shells. A shark's tooth was also found.

Between this and the next cutting to the south, a large pit on the east of the railway shows :—

Boulder Clay, 15 feet.
Sand, with a little gravel, 8 to 10 feet.
Red Crag, a mass of broken shells, 10 to 12 feet.

In the cutting beyond (southward) is coarse ferruginous sand, with many hollow nodules of ironstone.

Near Red Barn, W. of Marlesford Hall, are two pits in deep red, coarse sand, with ironstone. 200 yards N.N.E. of the barn this sand is overlain by light-coloured sand, gravelly at the base, with a sharp line of demarcation between the two.

About 50 yards N. of Marlesford Church is a pit showing the following section :—

Glacial Drift.
- Sand, with a little gravel.
- Fine, white, cross-bedded sand.
- Thin layer of gravel.

Decalcified Crag.
- Coarse, rather ferruginous, cross-bedded sand.
- Lenticular bed of nodular and laminated limonite.
- Very ferruginous cross-bedded coarse sand.

Frequent pits and temporary openings exhibit the decalcified sands of the Crag between Marlesford and Little Glemham.

Valley of the Alde, and Saxmundham Valley.

Highly ferruginous sand, probably referable to the Red Crag, is dug behind the school at Stratford St. Andrew, and near White Barn, between Great Glemham and Sweffling.

The well at the keeper's lodge in Dodd's Wood, on the left side of the river, is 45 feet deep, ending in what appears to be Crag sand.

A mile W.S.W. of Benhall Church, a sand-pit shows horizontally-bedded sand, over clayey sand with seams of limonite, over obliquely-bedded sand with hollow nodules.

Below Farnham Church is a pit in shelly sand, with two or more beds of wavy-laminated loam a foot or two feet thick.

The railway-cuttings east of the Alde, opposite Snape, are in the ferruginous sands with clayey bands: that at Rose Hill is the first to show a junction with the overlying Glacial sands. The sand-pit, half a mile east of the Snape junction, shows gravel over the ferruginous sands.

A pit, a quarter of a mile S.W. of the railway-bridge at Rose Hill, shows Boulder Clay (an outlier ? or Lower Glacial) over Crag sand.

At Benhall Lodge the well extends into the Crag.

Two hundred yards south of Benhall Church Glacial sand is seen overlying the decalcified Crag sand, whilst on the western side of the section the Boulder Clay sweeps abruptly down to the floor of the pit.

Half a mile E.N.E. of the church, a pit on the western side of the railway, hows the following section :—

Current-bedded sand and gravel, lying with a marked line of erosion on the bed below.
Brown Clay, two feet thick on the east, thins away west.
Horizontally-bedded Crag sand, with ferruginous bands.

At Saxmundham the Crag is over 100 feet thick, as proved by a well, at a point a few feet below the top of the formation (see p. 53).

A deep road-cutting, south of Sternfield, shows a band of grey clay above ferruginous sand, probably of Crag age. Similar sand is dug opposite Sternfield Hall, and also about 300 yards to the south-west, where the grey clay band again appears.

Valley of the Ore (or Alde), down the Left Side below Snape.

There are extensive exposures of the ferruginous sands near Snape Street, and they are touched in the pits at the brickfield, and at Rookyard Farm (see p. 29). Half a mile N.N.E. of the last place, the sand abuts against Boulder Clay by what appears to be an inverted fault, a view which is borne out by the undulations, to be recorded presently, of the lower beds of the Glacial Drift of these parts. That the Glacial period was one of oscillations of level is well known, and doubtless differential movements took place to a greater or less extent along lines of weakness. The section near Java Lodge, described on p. 16, bears evidence in the same direction.

The Crag sand is traceable up the lateral valley to Friston, and appears to be reached by a pit about 200 yards west of Lichfield House (north-eastward from the village) where the following section is exposed :—

> Boulder Clay, loamy to five or six feet from the top, stiff below; 13 feet.
> Alternations of sand and sandy loam, with fine gravel in places; the sand partly consolidated; thickness variable from resting on an eroded floor, but averages 4 feet.
> Tawny ferruginous sand (? Crag), horizontally-bedded on the east, but very obliquely on the west, with apparent contortions in the south-western corner of the section; 8 feet.

The sands are dug at Pound Farm (S. of Friston), and, on the edge of the marsh eastward, at Rushmere Farm. They are well seen at the spur named Cliff Plantation, and are dug in three or four sand-pits north and east. The clay at the brickyard, about half a mile west of Aldborough Station, is worked down to the sand; but no section of the latter occurs here. Owing to the rise of the Coralline Crag, the sands of the Red Crag here rapidly thin away to about 16 feet in thickness, but the protecting cap of Chillesford Clay has prevented complete decalcification, and shells occur throughout.

The outlier on which Aldborough stands has been still more protected (being cut off from lateral percolation from the mainland down to the level of the Coralline Crag, which division forms the isthmus of this quasi-peninsula), and a pit in very shelly Crag may be seen near the Water Tower (see Fig. 6).

Fig. 6.

Section in Ballast-pit, Aldborough.

Prestwich. *Quart. Journ. Geol. Soc.*, vol. xxvii., p. 335.

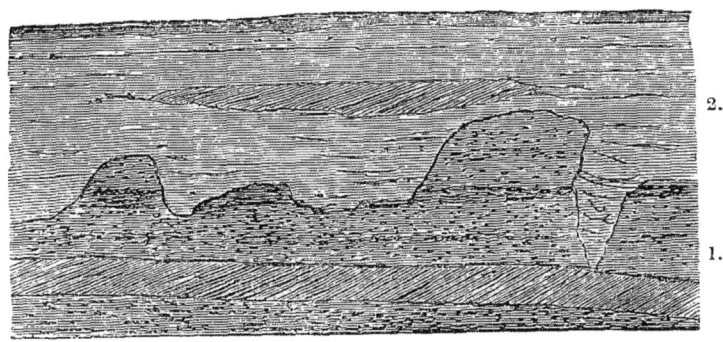

2. Brown Sands, with few shells and pebbles, about 10 feet.
1. Shelly Red Crag, about 10 feet.

Sea-board.

Shelly Crag is again seen on the high road above Aldborough Hall, and to the west of Watering House is a large pond by the roadside, apparently formed by the sinking of the Chillesford Clay into a large swallow-hole in the Coralline Crag. A little beyond the N. end of this pond is a pit showing at its southern face (1878):—

	FEET.
White sand (to west)	2
Boulder Clay with nodules of race at its base, lying erosively, with westerly dip, on the bed below	7
Chillesford Clay	5 on E. to ½
Sand passing horizontally and downwards into shelly Red Crag	4

Traces of shelly Red Crag remain in pockets on the surface of the Coralline Crag in the pit half a mile N. by W. of Aldborough Station. These occur on the northern side of the pit 40 to 60 yards from the road. Mr. S. V. WOOD, JUN., noticed here traces of the phosphate-bed.*

The railway-cutting east of this is sloped and turfed, but rabbits throw out shelly Crag, which is also traceable on the moor half a mile northward.

The Hundred River Valley and Minsmere Valley.

"S.E. of the cross-roads at Aldringham Green a pit showed gravelly sand (Drift) over white sand, probably belonging to the Crag."†

The ferruginous sands are seen under gravel in a pit 200 yards south of Aldringham Church and again at the farmyard to the east, where traces of the Chillesford Clay occur above the Crag sands. A similar section is given by a pit about a quarter of a mile south of the farm.

Half a mile E.S.E. of Aldringham Stone Cottage is the pit known in Crag literature as the Aldborough Thorpe Pit, situated on the southern edge of Aldringham Common. It is celebrated for its abundant well-preserved fossils, but shows no sub-divisions to be noted here. Traces of the Crag are seen over much of the common, and the flank of the hill extending to Thorpe shows occasional exposures of the same.

Between Thorpe and Sizewell the Glacial sands and clays extend below the sea-level, but at Sizewell the Crag rises well above the beach and the inland marshes, and is exposed in several pits. One of these, at the northern end of Sizewell Cliff, about 200 yards inland, is in very shelly sand (of littoral and estuarine character) mostly white, but somewhat ferruginous above, and containing many bands and lenticular masses of limonite. 250 yards from this is a pit in current-bedded yellow sand with clayey bands, and lines and stalactitic columns of limonite. A quarter of a mile south of this we again meet with white shelly Crag (due west of the word Furze on the map or N.W. of the Gap). Similar Crag is met with in Sizewell Gap and half a mile west of it, whilst at the farm a quarter of a mile S.W. of the flagstaff‡ the Crag, though shelly, is more ferruginous. Inland, towards Aldringham and Leiston Commons, the decalcified and oxidised Crag only is seen; but at the Leiston Ironworks well much shelly sand is brought up, proving that the decalcification, occurs only within reach of percolating, as distinguished from standing, waters. At a slightly higher level there is wavy bedded loam with sand-galls, possibly representing the setting in of the Chillesford Clay.

Goose Hill, N. of Sizewell, with perhaps the exception of its summit, consists of limonitic sand with occasional casts of shells.

The ferruginous sands continue to skirt the hill by Lower Abbey and East Bridge to Theberton and Middleton; but most of the outcrop is masked by the washing-down of gravel and bleached sand. The Crag sands are consequently seen only in pits made at that horizon, which are somewhat rare.

---

* *Ann. Nat. Hist.*, ser. 3, vol. xiii., p. 193.
† From W. WHITAKER'S Notes.
‡ As marked on the map. Erosion of the cliff may have necessitated removal and re-erection.

There is one on the New Cut in the outlier of gravel a mile from the sea, and another half a mile north-by-west from Theberton House. These and casual exposures have been our guides in tracing the Crag to Middleton Moor.

"On the northern side of Minsmere Level there is some difficulty in fixing the upper limit of the Crag, though it is a question of no practical importance. The Crag there consists wholly of unfossiliferous sand, but it is doubtful whether the topmost part of the sand here really belongs to the same formation as the rest, a question that may be better discussed in describing the tract just to the north, to which this area of some two square miles naturally ties on. Of course where one mass of unfossiliferous sand is overlain by another thin bed of like character it is practically impossible, and also practically useless, to draw a line between the two; at all events a satisfactory line."

"At the sand-pit, marked on the map, north of East Bridge, pebbly gravel and sand rests irregularly on sand, probably belonging to the Crag, and about a mile to the east another irregular junction of pebbly gravel with sand was seen."

"Shelly Crag has only been seen at one place, the foot of the cliff, where its occurrence was fortunately noted by Mr. E. T. Dowson and Mr. W. M. Crowfoot in 1871,* since which time the base of the section seems to have been hidden by fallen material."

"At the hillock named Coney Hill, at the junction of the marsh and the shingle, there is sand, probably Crag."†

CHILLESFORD CLAY.

This important bed, indicating the changes by which the shallow and turbulent sea of the Red Crag period became deeper and gradually less agitated by currents, has its southern limit practically coincident with that of the Crag at Walton Naze, Essex,‡ but is wanting, through denudation, over the interval between Walton and Butley, and has not been detected with any certainty westward of Wantesden (in 50 S.E.) and Halesworth (in 50 N.E.).

Whether its present limits in our district are approximately those of original deposition, as Mr. S. V. Wood held (1882), or whether here, as to the north, there has been extensive erosion in the Glacial period is uncertain.

The Chillesford Clay was first described under that name by Prof. Prestwich in 1849,§ and has since held its place in all classifications of British Tertiaries. It is a fine, almost impalpable, loam, generally grey, but often deeply stained with oxide of iron along the joint-planes and along the more porous beds. It passes down, in this district, almost imperceptibly into the Crag below, by gradual diminution of the proportion of clay, and by increase in size as well as in quantity of the particles of sand. But though this passage is so gradual in open section, the change from clay to sand is sharply marked in the soil, and the resistance of the tougher bed to rain-action has in many parts produced a palpable feature almost worthy of the title of escarpment.

The most westerly exposures of the Chillesford Clay of our district are in pits on either side of the road three-quarters of a mile west of Butley Church.

---

* *Proc. Norwich Geol. Soc.*, Part iii., pp. 80–83.
† From W. Whitaker's Notes.
‡ Whitaker. The Geology of the Eastern End of Essex . . ., p. 13, and authors there cited.
§ *Quart. Journ. Geol. Soc.*, vol. v., p. 345.

It is next seen in an old pit half a mile W. by N. of the church, after which its outcrop, though indicated by a slight escarpment, is concealed by sand washed down from the overlying Glacial beds.

In the outlier to the south it is seen in an old pit east of the road 500 yards from the church.

Where the outcrop changes its E.N.E. for a northerly course it loses the covering of sand, and is exposed almost continuously in ditches and old pits from a point half a mile N.E. of the church, by the east and north of a little wood, shown on the map, and then westward to the Oyster Inn, behind the stables thereof. An old pit 300 yards west, a ditch-section 150 yards beyond that, and traces of clay above the ferruginous sand on the eastern side of Staverton Park are the last remaining evidences, in this direction, of the Chillesford Clay.

The exposures of the clay at Chillesford Church and at the brickyard half a mile to the east, have been described in connection with the underlying Crag (p. 18). Old pits occur to the north and east and above the Decoy. Other pits, half a mile west-north-west of Chillesford Lodge and Sudbourn Hall respectively, indicate the course of the belt of clay between the Crag below and the Glacial sands above.

In the Orford outlier the clay is seen on the road over Broom Moor, at Sudbourn Church, and at and south-west of Ox House.

It is wanting for a space north of the outlier, but reappears on the east of Sudbourn Common, and, circling round Great Wood, extends to Webber's Whin and Calton Farm. Thence, with not infrequent sections, it forms the brow of the hill to Iken brickyard and above Redland's Covert, and west of Black Walks back to Chillesford.

The existence of an outlier to the north is demonstrated by a pit in the fields east of Tunstall Heath.

On the opposite side of the Ore the brickyard at the junction of the Maps shows about 13 feet of the clay, and a well, rather more than 16 feet deep, extends to the Coralline Crag. Sink-holes in the latter rock have produced several singular fallings-in of the clay, which descends below its true horizon, the bedding being thrown into festoons or curves, the upper parts of which are but slightly depressed, whilst the lower parts have been thrown into semi-circular or more complex forms. The exposure north of Watering House, described on p. 24, does not imply a thinning away northward of the Chillesford Clay, but a local erosion only; for in the brickyard three-quarters of a mile to the north-east we find about the same thickness of clay as in the southern pit on the edge of the map.

In the Aldborough outlier, the brickyard near the Water Tower showed 5 or 6 feet of ash-coloured clay with reddish bands, covered to an equal depth by sand, of which the lower foot or two feet was dark red and hard, probably from peroxide of iron accumulated at the plane of arrested percolation.

Shallow pits in Chillesford Clay may be seen half a mile E.N.E. of Aldringham Church, with a wash of gravel and sand above in the more northerly one.

A quarter of a mile west of Sizewell Boathouse is an old pit in yellow sand over grey clay with rusty joint-planes, probably Chillesford Clay, and the last trace of that bed in our district.

CHAPTER V.—GLACIAL DRIFT.

DIVISIONS.

THE Glacial Drift of this district consists of three distinct members:—

Upper Boulder Clay, with gravel in places.
Sands and Gravels, with seams of Boulder Clay.
Lower Boulder Clay, with seams of sand and brick-earth.

"Before describing these various members of undoubted Glacial Drift, it may be well, however, to notice a deposit of less certain classification, which has been mapped only over the small tract north of the Minsmere Level. This is a gravel, with occasional sand, composed for the most part of pebbles (chiefly of flint, but some of quartz); and whilst it seems to underlie the lowest beds of the Glacial Drift, rests generally irregularly on the Crag sand. Two of these irregular junctions have been noticed at p. 25."

"It has been given the name of Westleton Beds by PROF. PRESTWICH, whilst, somewhat earlier, MR. S. V. WOOD, Junr., classed it with his Bure Valley Beds; but in the Index of our maps it has been left unbracketed either with the Glacial or with the Pliocene Series."

"The possibility of its being represented, in a distant part of the London Basin, by a gravel of like character as well as of like doubtful age, has been alluded to elsewhere."*

"This gravel occurs to a greater extent in the tract to the north of our area, where it also caps Chillesford Clay."†

LOWER BOULDER CLAY.

This consists principally of a sandy loam with interspersed stones, mostly of small size; but occasionally the loam passes into, or is represented by, Boulder Clay of characters indistinguishable from those of the lenticular seams and masses in the overlying sands or from those of the Upper Boulder Clay. The bedding of the loams is generally well marked, and often much disturbed, whilst there is no evidence of this disturbance in the sands which follow. The irregular and spasmodic way, moreover, in which small outlying patches of these lower loams and Boulder Clays occur between the Crag below and the sands above, points to their former continuity as a general covering of the country, and to subsequent inter-glacial erosion in an interval apparently marked by the action of disturbing forces, resulting in a

---

* *Guide to the Geology of London.* Ed. 3, p. 57 (1880), and Ed. 4, p. 60.
† From W. WHITAKER's Notes.

complete change of the material conveyed to and deposited in this area. The characters, sequences, and general aspect of these lower beds agree closely with those of the First and Second Tills of the Cromer coast, with their intermediate beds, whilst the contorted stratification suggests a correlation with the overlying Contorted Drift of the coast.

Valley of the Deben.

The most southerly exposure of these beds within our district is at the brickyard west of Woodbridge, a quarter of a mile south-west of Farthing Cake Hall, where a long shallow section shows gently undulating beds of laminated loam, of varying shades of grey and buff, abruptly overlaid by the Upper Boulder Clay of the neighbourhood, and occupying the position, island-wise, of the sands and gravels which occur in full thickness close by, resting on the Crag. This is the Hasketon brickyard of Mr. S. V. Wood, Jun.

The Boulder Clay in the floor of the valley east of Java Lodge, Pettistree, is probably part of this series. As already described (p. 16), it shows a small fault, not affecting the overlying sands. The Crag sand has been partly torn up and re-deposited in alternations with the lower part of the clay.

The patch of clay westward of Java Lodge may be of the same age, as may the lowlying Boulder Clay east and north-west of Wickham Market.

Three-quarters of a mile north-north-east of Brandeston church is a pit in greyish loamy sand, with seams of clay, overlying Boulder Clay, the junction sloping sharply to the west.

A little Boulder Clay occurring on both sides of the valley between Brandeston and East Soham, and lying between the Chalk and the Glacial sands, may belong to this part of the series.

Valley of the Butley River.

The brickyard at the Rookery Farm, Eyke, shows, in the pits immediately west of Orphan's Piece, tough blue unstratified brickearth, like that of the Hasketon brickyard. Only about 6 feet is worked, but the deposit is proved, by a well, to extend some 60 feet down, and it consists of "black stuff all flaky like seaweed." This seems to indicate that the absence of lamination in the visible part is due to weathering. Traces of ferruginous gravel occupy "pots" and furrows in the surface of the clay, remnants doubtless of the sheet of gravel extending to the west. At the farm this gravel is over 20 feet thick, indicating a sharp descent of the clay bed. Southward of Orphan's Piece the clay has been dug in pits, but the one marked on the map is now a pond covered by reeds and bulrushes, and the others are ploughed over.

Valley of the Alde or Ore.

A considerable space is occupied by these beds on the north of Tunstall Heath, but only one section in them remains open, a pit three-quarters of a mile east of Tunstall Meeting-house, which showed, in 1879, the following:—

  Weathered chalky Boulder Clay, nearly 3 feet.
  Roughly bedded sand and gravel, 1 foot.
  Boulder Clay, an inch.
  Sand, passing down into the next bed, nearly 2 feet.
  Bedded clay sand and loam, 1 to 2 feet; passing into a wedge of
   Boulder Clay, 0 to several feet.
  Stony loam with streaks of Boulder Clay, over 3 feet.
  Sand in the floor of the pit.

In an old pit some three-quarters of a mile south-east of this, the light-grey finely-laminated loams peculiar to this series alternate, in an obscure section, with Boulder Clay.

On Iken Heath the Boulder Clay in the bottoms of two minor valleys clearly belongs, by its position, to this series, but the sections are very obscure, as is the case with the exposure of Boulder Clay south of Blaxhall Common.

Half a mile east by north of Blaxhall Church is a pit much obscured, but showing on its western face the sequence of laminated marly loam wrapping over the irregularities of a mass of reconstructed chalk, as in Fig. 7.

Fig. 7.

Section in a Pit half a Mile eastward of Blaxhall.

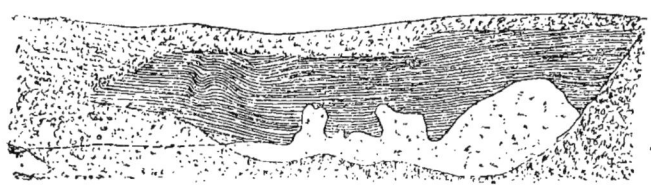

Part shaded by lines = loam.
Part lightly shaded = reconstructed chalk.
The rest, all round = overgrown parts, talus, &c.
The broken line roughly shows the continuation of the junction of the loam and chalk between the two exposed parts.

A quarter of a mile south-west of Snape church an old clay-pit, much obscured, shows sand, Boulder Clay and laminated loams. About a quarter of a mile eastward of this the following section was noted in a small pit:—

Yellow sand, 3 feet.
White, buff, and tawny laminated loam, dipping at 20° to N.E., 2 feet
Compact bedded sand with chalk-grains, 1½ to 3 feet.
Boulder Clay.

At the large brickyard marked on the map, about half a mile south of the high road, the pits are mostly in brown grey-banded brickearth, full, towards the base, of the calcareous concretions called hodmans (E. Anglian for snails = "race"); 20 feet of this is exposed, dipping southward at 5°. At a lower level Mr. S. V. Wood has seen "dirty-brown Boulder Clay, unstratified, full of chalk," resting on sand, and the southern end of the section showed a northerly dip. In the adjacent field to the south, he noted, in a pit, now ploughed down, "stratified clay with Boulder Clay intercalated in it, resting on sand, dipping at 30° ? to W."

On the east of the road two small pits show the base of the brickearth, resting on Boulder Clay, and this on tawny ferruginous sand, which we class with the Crag. The dip of the Glacial beds is here northerly, at a low angle.

At the Rookyard Farm is a continuation of the series seen in the brickfield, a large pit showing eight or ten feet of laminated brickearth, resting, with a slight northerly dip, on bedded sandy Boulder Clay eight feet thick, underneath which the tawny sands of the Crag are exposed. Half a mile north-north-east of this a pit shows a confused mass of brickearth, sand, and Boulder Clay faulted against Crag sand, the fault-face sloping at a high angle southward (in reverse of what is supposed to be the normal relation of hade to downthrow). All the bedding-planes dip in the same direction.

A trace of the Boulder Clay of this series is seen at Friston, but it is very thin, whilst eastward of Pound Farm, to the south, it is seen by pits to retain its full thickness. Its base is shown over Crag sand on the north of the Decoy. It is exposed in casual sections between Park House and Rushmere Farms, but is lost southward by thinning away, whilst to the north and east it is covered by higher beds.

Northward from Aldborough.

This division seems to be represented by the Boulder Clay, already noticed, in the pit behind the Watering House, see p. 24.

A bed of yellow clay, full, in places, of chalk nodules and containing some large flints, was exposed in a pit a furlong west of Aldringham Green, and was noted by Mr. S. V. Wood, but no trace of the pit can now be found. The bedding undulated in wide curves, and in part of the pit the clay was covered by compact yellow gravel.

A large overgrown pit half a mile W.S.W. of Sizewell flagstaff showed in July 1878 the following complicated sections:—

	S. part.	Western side.	N. part.
	Boulder Clay.		Chalky gravel.
	Grey loam.		
	Gravel.		

	W. part.	Northern side.	E. part.
	Boulder Clay.		Chalky loam.
			Red sand and gravel.

	N. part.	Eastern side.	S. part.
	Boulder Clay.		Chalky loam.
	Hard red sand and loam with pebbles.		Red sand (? Crag).

Southern side. E. part.
Boulder Clay, 10 feet, irregularly on fine white sand.
Red sand abutting against gravelly sand with loamy seams.

"A quarter of a mile northward of this another old pit (marked on the map), gave, at its one clear part, the following section, the beds being of a like character to those of the above, and the Drift filling a hollow in the Crag."*

Glacial Drift. { Laminated clay.
Gravel and sand, thin.
Boulder Clay.
Sand, probably Crag.

SAND AND GRAVEL.

Sand and Gravel.

Though these beds extend over a very large proportion of our area and form the actual surface of a not inconsiderable fraction of it, there are few sections exhibiting notable features or deserving more than a passing allusion. Such as only consist of the normal white sand reached in the floor of pits sunk in the Boulder Clay will be mentioned in our account of that division. The bulk of the deposit is fine white sand horizontally-bedded, but there are masses of gravel in places, and sometimes of Boulder Clay, in more or less wedge-shaped sheets. The gravels are the less sought after because the habitual collection of stones from the Boulder Clay fields supplies abundant road-metal, and where the Boulder Clay is absent it is carried, in preference to gravel, for the repair of roads across the sandy heaths in the south and east of our district, furnishing at once stone, calcareous cement and clayey matrix.

Valley of the Finn.

Two large sand-pits occur at the fork of the roads half a mile north-west of Tuddenham Hall.

---

* From W. Whitaker's Notes.

A quarter of a mile south-south-west of Culpho a pit shows the following section:—

		Feet.
Glacial Drift.	Gravelly sand and soil	6
	Grey clay with seams of dark red and grey sand	4
	Reddish-yellow sand, with a thin layer of coarse gravel at the base	from 1 foot to 3
? Red Crag.	Current-bedded yellow sand	8
	Tawny sand	3

Eastward of Culpho Church the sands are also somewhat gravelly, and gravel caps the Crag in a pit east of Grundisburgh Hall. This section and that west of Grundisburgh Church are described on p. 14.

Laminated clayey sand is dug in a pit at the head of Otley Bottom, half a mile south-south-east of the church, and shows a passage upward into the Boulder Clay. Red sand and gravel are exposed, at the top of the series, in a pit at the farm half a mile north-north-west of Clopton Church, under the Boulder Clay. Gravel is raised in a large pit south-east of Hasketon Parsonage.

Valley of the Deben, up the Right Side.

Whitish sand and gravel, with worn and broken shells, is seen under Boulder Clay in the lane about half a mile E. of Farthing Cake Hall (N.W. of Woodbridge). The shells appear to be derived from the denudation of the Crag, the fragments occurring more abundantly near the outcrop of the Crag. West of Fern Villa gravel is dug.

There is a large pit in fine sand about 600 yards east of Bredfield White House, N. of Woodbridge, and a more complex series in the grounds of Foxburrow Hall has been described on p. 15.

Gravel is dug largely between Melton and Ufford and west of Ufford Street.

Near Byng Hall, as already mentioned, p. 16, the Crag sands rise within a few feet of the Boulder Clay, so that the width of outcrop of the Glacial sand at this point does not imply corresponding thickness, but merely a superficial draping of a slope of the Crag surface.

Chalky gravel, with an irregular seam of marl, is seen under Boulder Clay in a pit 200 yards south-west of Mountains Barn.

A little patch of Glacial sand overlaid by Boulder Clay rests on the slope of the Crag sand north of Byng Hall, the clay pit marked on the map showing all three deposits. On the road to the east of this there is gravel.

Another patch of Boulder Clay, but apparently one of the lenticular masses frequent in this series, occurs on the high road about half a mile north of Ufford Street. In the adjacent field to the east gravel is seen overlying dark red Crag sand, and a similar junction occurs in a pit by the railway a mile to the eastward. Gravel is also seen half a mile north of Loudham Hall.

The sand-pit three furlongs south-east of Wickham Market church shows traces of Boulder Clay at top, and the old brickyard north of this is in Boulder Clay, the bottom of which is not reached.

Near the river, half a mile N.W. of the church, a large pit shows on its north-eastern side 10 feet of Boulder Clay, with some coarse gravel above, which, though bedded, may not be in place, as spoil-heaps and talus show lines of varying material, simulating bedding. The western part of the pit gave the following section:—

Coarse gravel.
Loamy sand with patches of gravel and seams of grey clay. A few fragments of shells were noticed. The upper part is tufaceous from infiltration. A plane of erosion separates this from the bed below.
Gravelly sand, the gravel in thick wedge-shaped masses, the sand in seams at high angles of deposition, crowded with flakes and large angular slabs of Oolitic shales. Traces of shells are scanty, a fragment of a *Cardium* being the only recognizable piece.
White sand.
Fine gravel.
Coarse open gravel consisting largely of chalk pebbles.

Boulder Clay is seen in the road above, but its relation to the sands and gravels is unknown.

Gravelly sand is seen in the Charsfield tributary-valley and is dug near the Gull and Park Farms.

There is also gravel eastward of Letheringham Old Hall.

At Hoo, about 600 yards west of the church, is seen Boulder Clay passing down, by alternations of chalky gravel and fine clay, into a thick lenticular mass of similar clay, resting on sand and gravel, in which occur other seams of clay.

Westward of Cretingham and northward of Framsden Hall gravel prevails and is dug in one or two places.

In the side-valley between Pettaugh and Winston exposures are somewhat scanty. Sand and gravel are seen over the little patch of Chalk, and in an old pit, east of the road, fine gravel and sand are separated from the Boulder Clay by a loamy passage-bed. West of the road, on the northern side of the hollow, is a pit in white sand, becoming very loamy and well-laminated towards the top, and divided by a thin bed of brown loam from the Boulder Clay.

Similar loam in a like position is seen at the brickyard half a mile north of Winston.

Valley of the Deben, down the Left Side.

Gravel is raised on both sides of the road about a mile north of Debenham, and pebbly beds occur in the sand southward of White House Farm.

In the Soham tributary-valley a little gravel is seen near the bridge south of Cretingham Manor Farm, and opposite King's Hill. A small gravel-pit was open in 1876 at Clow's Corner, and gravelly sand extends up the valley to Monk Soham. West of Earl Soham Lodge again gravel is dug to a slight extent. South-east of Earl Soham Church a deep road-cutting shows sand with seams of grey clay, and higher up there is Boulder Clay, but the junction is not exposed. The sand is dug on the west of Brandeston, half a mile south-south-west of Hill Farm.

Sand also skirts the river in the grounds of Brandeston Hall, whilst the inlier north of the road is principally gravel.

In Kettleburgh parish sand skirts the Deben continuously, and is seen under the clay in the lateral hollow occupied by the village: to the east of this a good deal of gravel occurs in the sand. The fine pit near the western side of Easton Park is described on p. 17. The hill on which Lady Hamilton's Model Farm stands is mostly gravel.

There is a large inlier, consisting chiefly of gravel, in Easton Park, and a second and much smaller one in the village to the south-east, where, at the clay-pit marked on the map, is seen about 16 feet of coarse brown and grey sand, with seams of coaly detritus and waterworn fragments of shells. The overlying Boulder Clay is 12 feet thick at the northern end of the section, rising to nothing at the southern end.

The crooked lane shown on the map south of Rookery Farm, Hacheston, is really a deep watercourse, with banks of gravel and sand, covered by loamy rainwash.

A small lenticular mass of Boulder Clay occurs on the high road east of Wickham Mill, and traces of it are seen in the lane on the south.

North of Ashmere Hall an old pit shows the base of a thin bed of Boulder Clay, which is traceable thence to the Crag pits half a mile southward, see p. 17. To this belt of calcareous clay is due, in all probability, the consolidation of the upper layers of the sand, in a pit south-west of Ashmere Hall, by the deposition of tufa in the interstices between the grains.

In the railway-cutting to the south the seam of Boulder Clay is again seen on either side. It is now from three to six feet thick, and lies (on the northern slope) at an angle of 30°, but less on the south. The dip is eastward.

There is some room for doubt as to the age of the spread of gravel south-east of Eyke, its superposition to the Boulder Clay seen at intervals north and west of it being very uncertain, and we only know that it is newer than the loam of the Rookery brickfield, so that it may belong to the Glacial gravels or may be an exceptional form of the deposits of the succeeding Upper Boulder Clay period. Its boundary consequently is shown on the map though it is coloured uniformly with the sand to the south.

Valley of the Butley River.

Only two sections within the area drained by this stream and its tributaries call for mention here.

Half a mile north of Ivy Lodge, Rendlesham Park, an old pit, now ploughed down, showed, in 1876, under a very irregular covering of Boulder Clay, a varying depth of sand, with two thin layers of bedded Boulder Clay separated by stony loam. Half a mile northward of this is a sand-pit with three or four seams of Boulder Clay from two inches to a quarter of an inch thick, and about an inch apart, coalescing into one bed at one side of the section.

Valley of the Ore, up the Right Side.

Coarse gravel is raised half a mile eastward of Blaxhall Church, and sandy gravel ranges between Blaxhall Street and Beversham Bridge.

Sand is dug in two places in the Deer Park, Campsey Ash, and in a pit about 200 yards north-east of Campsey Ash Church, where it is intersected by a band of Boulder Clay.

About three hundred yards westward of the junction of the Framlingham branch with the main line of the railway coarse chalky gravel is seen.

There are old gravel-pits on both sides of the high road half a mile westward of Marlesford Station. On the road running parallel with the railway about half a mile N.W. from the station, a thick lenticular bed of Boulder Clay occurs in gravel and sand. In the road-section sand is exposed below it, and a large old pit on the east of the road cuts through it into sand. On the west of the road, at the fork, is an old pit mostly in clay but very gravelly on the southern side. A large piece of red chalk was found in this clay.

In the gravel-pit about 50 yards west of this a streak of Boulder Clay, 6 inches thick and 3 or 4 feet long, was exposed in 1876. The floor of the pit seems to be Boulder Clay, probably the bed seen in the road. The gravel, of which over 20 feet is seen, contains many pebbles of chalk.

Northward of Hacheston Church gravel extends to the further end of the village and up the hollow by Bloomville Hall.

The cutting on the railway opposite Broadwater (south of Framlingham) is in sand, with a little gravel. At a point about 130 yards west of the 89th milepost, is earthy gravel with a seam of clay showing strong contortions and even inversion.

Sand and gravel occur on the railway at the ticket-platform at Framlingham, and the brickyard to the east showed, in 1876, 8 feet of yellow and brown stony loam, with seams of gravel and sand, overlying loamy clay.

The irregular surface of the sand under the Boulder Clay at Framlingham causes no less than three inliers of the sand, above and below which the Boulder Clay extends to the bottom of the valley. Two of these lie west of the town; the third is on the north, partly concealed by the alluvium of the Mere. The most westerly consists of gravel and sand, the next of sand only.

A well at the Almshouses, south-west of the Mere, showed a little Boulder Clay at the top, and another bed, about a foot thick, 16 feet down, with sand above and below.

Valley of the Ore, down the Left Side, including the Valley of the Alde.

The railway-cutting, nearly a mile S. of Framlingham Station, passes at its northern end through a ridge of sand and gravel concealed by Boulder Clay. This ridge showed, in 1875, the following section on the western side of the railway :—

Bedded sand and fine gravel with much chalk.
Lenticular mass of coarse chalky gravel with a boulder of Boulder Clay.
Bedded sand with more and coarser gravel than the upper bed, and boulders of Boulder Clay.

The northern side of this section presented a nearly vertical face, against which Boulder Clay abutted.

On the eastern side of the railway, at the top of the bank, was bedded chalk-gravel and Boulder Clay, the relation between them being obscured by talus. Lower down and beneath the Boulder Clay, which abutted against a very steep face of gravel, was the following section :—

 4 Bedded clayey sand.
 3 Coarse gravel, rather ferruginous.
 2 Bedded clayey sand.
 1 Ferruginous sandstone; a piece of shell (? Crag).

1 and 2 had a slight northerly dip, but 4 occupied the singular position (for such a bed) of resting against a face of 3, inclined at about 60° in the same direction. These beds were sharply cut off, horizontally on the top, and on the southern side by a face curving from the vertical to 45°, the whole being enveloped, as on the western side, by Boulder Clay with large tabular and unworn flints. The chalky-gravel at the top of the cutting may be part of the Boulder Clay series. It is seen in the road close by.

North-west of Broadwater is a large pit in a plantation, mostly showing white sand, with a little chalk-gravel, and some masses of reconstructed Chalk. Boulder Clay is seen at the northern end of the plantation.

Sand has been dug in a large pit at the western corner of Parham Wood.

In the cutting W.N.W. of Parham Hall a steep-sided channel has been eroded in the Crag sand, and lined with sandy gravel, a mass of Boulder Clay occupying the centre of the hollow.

In the valley of the Alde proper a very fine section is afforded by a pit about a quarter of a mile south of Coldstone Hall, above Bruisyard, where the following beds are seen :—

Boulder Clay	3 Feet.
Fine white current-bedded sand	13 ,,
Ferruginous gravel	1 ,,
White gravel	2 ,,

A pit at the back of Hazlewood Hall, S.E. of Friston, shows the following sequence at the top of the series with which we are now dealing :—

Boulder Clay, 5 feet.
Chalky gravel, 1 foot.
Brickearth with chalk, passing, by alternating laminæ, into the bed below, 1½ feet.
Yellow sand, 2 feet.
White sand with scanty shell-fragments, 2 or 3 feet.
Reddish fine gravel.

Valley of the Hundred River.

At Billeaford Hall, E. of Friston, the top of the series is seen in a large pit to be as follows :—

Boulder Clay, with an uneven jagged junction, 4 to 5 feet.
Laminated loam with pockets of white sand, passing down into the next, 10 to 12 feet.
Fine white sand.

A quarter of a mile westward of this, a somewhat different junction is seen, viz. :—

	Feet.
Boulder Clay with sand-galls	4 to 5
Yellowish sand	1 to 2
Boulder Clay	1
Brownish-yellow obscurely-bedded loam with bits of chalk, passing down into sand	12

"A like section was noted about an eighth of a mile N.W., showing Boulder Clay, over sand and Boulder Clay, over sand."*

---

* From W. Whitaker's Notes.

Coarse gravel has been dug at Coldfair Green and Aldringham Green. On the Common to the east of the latter the sand is blackened by peaty matter to five or six feet deep, below which it is a clear bright yellow.

Consolidated flint-gravel occurs on the cliff a quarter of a mile south of the building delusively called on the map the Tea House.

Valley of the Leiston Brook.

In a large brickyard, a quarter of a mile northward of the railway-station, a fine section was exposed in 1876, as follows:—

Yellow sand, 8 to 12 feet.
Brown laminated clay, varying from 0 to 6 feet.
White current-bedded sand, with seams of race, septaria and clay nodules. (The bedding dips N. at 30°), 3 to 2 feet.
Well-laminated greyish clay, with pockets of coarse sand, over 10 feet.

The section was somewhat altered in 1878 and showed the following:—*

False-bedded light-coloured sand, ? partly gravelly, up to about 12 feet.
Grey laminated loam, with layers of sand; some broken shells in a patch of coarse sand with small pebbles; nearly 15 feet shown.

At the other (E.) part of the pit the upper bed was hardly shown, but there was a gravelly soil over the lower bed, here consisting of sand with layers of grey laminated clay.

The same (lower) clay is dug in an inferior section, in a brickyard 200 yards east of the station, and it has been traced in the valley to the south.

Minsmere Valley.

The stratified sandy clay passage-bed at the base of the Boulder Clay is well seen in pits E. and N. of Upper Abbey Farms, north-eastward of Leiston.

Opposite Theberton Hall, on the eastern side of the high road, a bed of contorted chalky gravel, with seams of sand and loam, occurs at the base of the Boulder Clay, and may be regarded as belonging to either division of the series.

In the side-valley south of the Plough Inn, S.E. of Middleton, there is a good deal of pit gravel. Higher up, at a pit on a by-lane half a mile west-north-west of Theberton Hall, a bed of irregularly-laminated brickearth, with grains of chalk, separates the sands and the Boulder Clay.

"On the northern side of the valley, in a long old overgrown pit in a field about a third of a mile W.N.W. of Scott's Hall, there is, at the eastern end, some Boulder Clay, and near the bottom some gravel, cemented into hard stone by calcareous cement. At the middle part there is a good deal of sand; and at the western part there is at top, on the southern side, loam full of stones (= weathered Boulder Clay) with a trace of Boulder Clay in it and sand beneath.

"Another old pit, at the edge of the next field westward, is in like beds, with the hard conglomerate underlying the Boulder Clay."*

---

* From W. WHITAKER's Notes.

CHAPTER VI.—GLACIAL DRIFT—(continued).

UPPER BOULDER CLAY.
General Account.

THE greater part of this district is covered by a sheet of clay, more or less charged with round and sub-angular stones, largely chalk and flint, but with a considerable proportion of older rocks. There is every gradation in size, from the smallest particle up to blocks weighing over a ton, but it is not common to find stones exceeding 10 lbs. in weight.

The clay varies in colour and texture within wide limits, from compact blue stony clay to light-yellow porous loam with few or no stones. Sometimes it changes, by deficiency of matrix, to a mere aggregation of stones and sand, available as gravel, but such is not very commonly the case.

It has been extensively used as top-dressing for light land, and the more calcareous parts for such of the heavy land as was deficient in lime-compounds It is used, as has been already noted, to repair the roads on the sandy heaths, and the stones picked off the fields serve as metal for the thoroughfares in the clay-districts. Though scantily used for brickmaking, it is employed in building as "clay lump," being worked with straw into blocks about 18 inches long by 9 inches square in section. When thoroughly air-dried and set in moist clay, these form substantial walls, less permeable to moisture than brick, and only requiring timber-framing to support upper floors or roof. Houses thus built indeed are generally drier than those of more durable (?) material; for, with judicious protection against direct injury by weather, the clay will outlast the timber.

Valley of the Finn.

The base of the clay was observed about a hundred yards south of Witnesham Church, and in the road-cutting south-west of Tuddenham Hall.

A quarter of a mile east of Culpho Church the sand and gravel of the small tributary hollow are seen to pass under the Boulder Clay.

Up the Otley branch of the valley, in a roadside-pit near Grundisburgh Hall, a somewhat similar section occurs. Eastward of Grundisburgh the clay descends to the brook and so continues to Clopton, above which, on both sides of the valley, the base rises slightly, exposing the underlying sands. These are seen half a mile south-east of Otley Church, and the boundary of the clay skirts the hillside to the Crown Inn, when it descends to the valley again for a few hundred yards. It rises again at the farm half a mile north-north-west of Clopton Church, where a pit shows red sand and gravel under the Boulder Clay. The clay again descends between Clopton and Burgh, but rises at the latter village, and the base, skirting the hill, is exposed in a pit south-east of the church, and again about a hundred yards west of Burgh White House. From near this to Hasketon the bottom of the valley is occupied by the Boulder Clay, whilst on the flanks on either side protrude inliers of the Glacial sands and Red Crag.

The Boulder Clay lies indifferently on the Glacial sands or on the Crag, the sands being sometimes absent or very thin : thus they are absent at the pit a quarter of a mile east of Hasketon Hall, and are very feebly represented the same distance further eastward, and near Bealings House, though fairly developed between these points. Similarly at Great Bealings, patches of the Boulder Clay are found, some on the Glacial sand, there forming a continuous sheet, and one on the slope of the Red Crag below. These data show that the Glacial sands were either deposited in a very irregular manner as to thickness and continuity, or were extensively denuded before the deposition of the Boulder Clay. I incline to the former view, as more consistent with the facts afforded by examination of the whole district. A series of sand- and shingle-banks, such as must have constituted the bed of the sea in which these Glacial sands were formed, must have had many channels ramifying hither and thither, often reaching to the lower part of the Red Crag beds, and sometimes to the London Clay or to the Chalk. The infilling of these channels by clay, would necessarily produce apparently capricious risings and fallings in the base, of the clay, as exposed by modern denudation. In the valleys of the Deben and Ore similar phenomena are presented.

The singularly-shaped inlier of sand north and east of Hasketon offers but one junction with the overlying clay, in a pit about 200 yards to the north-east of the church.

Valley of the Deben, up the Right Side.

The section at the Hasketon Brickyard has been already described, p. 28. A quarter of a mile east of this Boulder Clay is seen, resting on sand, 300 yards south of Farthing Cake Hall, and half a mile to the east, beyond a lane in a deep gully, an old clay-pit occurs on the point of the promontory of Boulder Clay.

The outlier east of Bredfield White House is penetrated by a pit reaching sand ; a much smaller outlier is seen to the east.

The pit west of Foxburrow Hall has been described on p. 15.

West of Ufford Place large old pits show the base of the clay, which is also exposed at three points near the entry to Byng Hall (to the north).

On the eastern edge of the inlier of sand, in the valley S.W. of Dallinghoo, are no less than four pits showing Boulder Clay over sand. The most southerly of these is on the by-road to Bredfield; another is a hundred yards to the north, the third directly below High House Farm, and the fourth near Moat Farm : the last shows six feet of Boulder Clay and 16 feet of sands.

The exposure of the base of the clay near Mountain's Barn and that of the little outlier below Pettistree Green have been described on p. 31. Other outliers deserving only bare mention occur, (1), half a mile south-south-east of Java Lodge ; (2), at Loudham Hall ; (3), on the railway eastward of the Hall, and (4), a quarter of a mile north of the Hall.

The base of the clay is seen in pits near Galham Hall (Wickham Market), and, up the Potford tributary-valley, at Park Corner House, below the Gull Farm, and near Charsfield Hall. There are also several junction-pits between the Meeting House and Letheringham Lodge, below the last place, and in a large pit half a mile to the east.

The mode of occurrence of the Boulder Clay in the upper part of the Deben valley is less regular than in the area drained by the Ore and Alde ; or perhaps it would be more correct to say that, in the Deben area, the surface of the Glacial sand is much less even, its undulations not coinciding with existing valleys, so that the base of the overlying Boulder Clay rises and falls in an apparently capricious manner.

From the pits near Letheringham Lodge the clay sweeps down to the level of Potford Brook, and slowly rising again, skirts the Deben at a few feet above the alluvium for nearly a mile, when it rises sharply to cross the ridge, and descends to the river again at Letheringham Old Hall, for the most part resting directly on Red Crag. It is probably continuous under the river to Easton. Rising to the west, it passes Letheringham with fluctuating elevation, to come down to the Deben again directly north of that village, and to occupy that position throughout the north of Hoo parish (except for a narrow outcrop of sand on the north-east), whilst the sand, as shown by the smaller valleys, forms ridges between Letheringham and Framsden on the south-west, and Hacheston and Earl Soham on the north-east. At Brandeston, and between Friday Street and Cretingham, the clay is at a low level, though probably not below the alluvium.

What other depressions may be concealed under the broad spread of the clay to the north and west it is impossible to say, but the valleys appear to indicate a fairly uniform course beyond Earl Soham and Ashfield.

Immediately to the west of Letheringham Old Hall the base of the clay rises, and is seen in a pit 700 yards N.N.E. of Letheringham Lodge. A similar section occurs 200 yards south-west of Letheringham Church, and sand runs up the valley to near Goodwin's Place. Another junction is seen about 400 yards west of the church, and a fine section in a pit by the roadside north-west of the church, where the upper part of the sand, for some two or three feet from the clay, is consolidated into a soft rock by carbonate of lime, carried by percolation from the overlying Boulder Clay.

Up the valley, between Hoo and Monewden, the base of the clay rises considerably, and west of Hoo Church a pit shows the junction with the underlying beds. This is again seen in another pit some two hundred yards to the south, and yet again in a run-down pit on the further side of the road to Monewden.

The junction is next seen a quarter of a mile S.S.W. of Monewden Hall; whence the boundary-line gradually descends to the wood, and back to near the Hall, where it rises sharply northward, and again descends (north-eastward) to the farm in the bottom. North of this it crosses a low ridge of sand, and runs, at a small height above the alluvium of the Deben, to Friday Street, beyond which, for some 300 yards, it is at or below the level of the river. The sand comes out again after this interval, but only as a narrow belt, and as an inlier formed by a gully a quarter of a mile below Cretingham.

The sand-pits on the road between Cretingham and Framsden are no longer good sections, but both originally showed the base of the clay. In the absence of such sections it is uncertain whether the sand in the hollow between Pear Tree Farm and Welham's Grove is of Glacial age, or a more recent deposit, from the denudation of the Boulder Clay. A similar lack of sections obtains in the branch-valley that supplies and drains Framsden. It may be remarked that the natives are in no wise particular as to their water-supply, but it is questionable if stagnant ponds are not safer than open brooks, or even than wells in coarse loose gravel, that transmits, with little filtration, the washings of cultivated land and of spaces inhabited by people whose sanitary science is *nil*.

Gravelly rainwash covers the Boulder Clay along the northern side of Helmingham Park, and the clay is weathered to a depth of more than 10 feet at the brick-kiln eastward of Bocking Hall N.W. of Helmingham where the resulting loam is used.

In the side-valley, between Pettaugh and Winston, the only exposures of the base of the Boulder Clay have already been described. The junction is obscured by grass in the old brickyard at Barley House Farm, east of Winston, but was visible (1876) near the Malting Farm, half a mile to the west, and at the brickyard north of Winston village.

Valley of the Deben, down the Left Side.

Junction-sections are scanty about Debenham, but there is a pit about 300 yards south of White House Farm, showing Boulder Clay over pebbly sands, and at Thorpe Hall, west of Ashfield, exposures of Boulder Clay and of sand were observed in close proximity, but the actual junction was not laid bare, nor is it again seen for about a mile, when an old pit west of Wood Farm shows the junction pretty fairly. The base descends thence to the Deben, along the valley of which, to Cretingham, it remains in this low position, though rising to the north, as shown by the side-hollow east of Wood Farm.

This slope of the base is again proved a quarter of a mile northward of Cretingham, where a road-cutting forms an inlier of sand under the Boulder Clay, although the latter descends the slope nearly to the river. At the Farm eastward there are pits in clay, to the south-east of the house, but to the north is a large sand-pit with very little clay cover.

The junction is again seen up the Soham Valley at Manor Farm, and, running pretty evenly round the hollow by Cretingham Lodge, the base of the clay descends to the stream at Earl Soham.

Half a mile south-south-west of Hill Farm is a pit, dug for sand through the clay, which descends sharply to the south, but shows a large inlier above the road west of Brandeston church. To the south-east the clay reaches the alluvium of the Deben, but the sand rises up in the hollow between Brandeston and Kettleburgh.

North of the road and east of the brook is a sand-pit, at the back of some farm-buildings, showing a mere trace of Boulder Clay at the top. The clay descends to the road for a short distance, and afterwards skirts the hillside above it, a large junction-pit occurring halfway between the brook and Kettleburgh. Similar sections are seen on each side of the hollow in which the village of Kettleburgh lies, and a quarter of a mile east of the water-mill.

The base of the clay is seen in a junction-pit in the fork of the brooks west of Martley Hall (Easton), and thence it descends to the Deben. Rising again almost immediately it skirts a hill of gravel, reaching the river again near the corner of Easton Park. North of this the underlying sands are reached through the Boulder Clay in the pit described on p. 17 and in the large inlier in Easton Park.

The presence of an inlier of sand in Easton village denotes a second boss. In the pit described on p. 32, the northward fall of the Boulder Clay base is exposed, and on the west of the lane another pit shows a still sharper descent, there being 15 feet of clay on the northern side of the face and none on the southern. Two small outliers to the south reach down the slope nearly to the level of the alluvium.

The small exposures of Boulder Clay between Easton and Glevering Hall are probably in outliers, separated from the main mass by denudation.

The junction is seen in an old pit about 400 yards S.S.E. of Glevering Hall, and again on the east of the park, besides casual exposures. In the inlier between this and Hacheston a large pit, half a mile N.E. of the hall, shows the base of the Boulder Clay.

A quarter of a mile south of Rookery Farm (Hacheston) an old pit, by the roadside, shows Boulder Clay to the north and sand on the south, with a steeply-sloping line of division. About 350 yards south of Beggars Barn is a large junction-pit, and a similar section, much obscured by weathering, may be seen by the side of the high road. Other exposures of the base of the clay occur at 500 yards to the S.S.W., the same distance to S. by E. and a quarter of a mile N.E. of Ashmere Hall (west of Campsey Ash).

The base is next seen on the railway south-west of Campsey Ash station. Thence it crosses to Copperas Wood and, turning westward, is seen in a pit on the western side of the road to Rendlesham, and again a quarter of a mile north-west of Rendlesham House.

A tiny patch of Boulder Clay is seen capping sand in a pit a quarter of a mile southwards of Rendlesham High House, and in the stack-yard at Naunton Hall a mass of Boulder Clay, with almost if not absolutely vertical sides, has been left by the excavation from around it of the Crag, in which it must have occupied a deep gully, the trend of which was N.N.W., at right angles to the course of the Deben at this point, but parallel to the upper part of the valley and almost pointing up it.

Traces of Boulder Clay are seen at intervals along the road to beyond Eyke, and these are perhaps in a continuous mass with a small exposure about three-quarters of a mile east of Bromeswell Church. The boundary of the Eyke and Rendlesham patch is much obscured by drifting sand, but the clay extends eastward, and is seen among the cottages north of Orphan's Piece, and in pits between Tithe Cover and Rendlesham Red House.

Valley of the Butley River.

The Boulder Clay of the last-mentioned sections, on the east of the watershed between the Deben and Butley Rivers, is separated only by an accident of denudation from a spur descending from the main mass near Rendlesham Red House. This spur is merely a slight hollow in the sands, in which the clay rests, as it does in the lower hollow of the outlier; indeed it is possible that the clay does extend down the intervening slope, masked by drifted sand.

The Boulder Clay, extending from Rendlesham to Wantesden, is very thin, and is pierced in many pits: its surface is also so much covered by sand, blown from the adjacent heaths, as greatly to obscure its limits. Clay is seen in junction-pits south, west, and north of Wantesden Church, at 350 yards, and again at half a mile, south of Ivy Lodge.

There is a small outlier, proved only by a pit nearly as big as itself, a mile westward of Wantesden Hall.

Half a mile to the north of Ivy Lodge, Rendlesham Park, a pit, now ploughed down, showed, in 1876, a very irregular junction of the sand with the Boulder Clay. On the north-west the base of the clay was at the top of the section; on the west, south, and south-east at the bottom; on the east-south-east it rose suddenly beyond the face of the section; it descended to form a thin patch on the east, and then came up again; it reappeared at the north plunging rapidly down, but rose again immediately to the north-west. With so irregular a junction, it is obvious that boundary-lines are merely generalizations.

Half a mile north of this is another old pit showing the junction. Similar sections are afforded by a pit half a mile west of Tunstall, and by another half way between Tunstall and Ivy Lodge. A trace of clay is also seen capping the sand in a pit on the road south-east of Potash Farm.

Between this and Chillesford are over a dozen pits showing clay over sand thus:—

(1.) Half a mile E.N.E. of Potash Farm.
(2.) Marked on map, east of Cats Grove.
(3.) 300 yards N.E. of (2).
(4.) 500 yards S.E. of (2) and a quarter of a mile W. of (5).
(5.) East of road half a mile southward of Meeting House.
(6.) 500 yards S.E. of (5).
(7.) 200 yards S W. of (6).
(8.) 600 yards S. by E. of (6).
(9.) 300 yards S. of (8).
(10.) 250 yards E.S.E. of (9).
(11.) Back of Chillesford Church 300 yards S. by E. of (10).
(12–14.) Two pits marked on the map on the east of the clay spur, with another between them.

Mr. S. V. Wood, Junr., noted in No. 10 (now much run-down and showing only the Boulder Clay, and that very obscurely) that faint planes of bedding in the clay were comfortable to those of the sand, both being slightly arched.

In No. 11 the Boulder Clay is similarly bedded in the hollows of the Chillesford Clay, on the eroded surface of which it rests, but shows no trace of bedding in the upper part.

Valley of the Ore, up the Right Side.

Pits Nos. 12–14 of the above list are on the eastern side of the spur of Boulder Clay, and therefore within the Ore Valley. There is a fine junction-pit rather more than a quarter of a mile east of Tunstall Church.

At a house about 400 yards N.W. of the church a well showed loam and red sand in irregular masses, penetrated by more or less vertical lenticular veins of blue Boulder Clay of normal character. Similar material occurs along the road to the north-east.

North of Limetree Farm, north of Tunstall, is a junction-pit of the usual type, and others occur on the road at a quarter and at half a mile southward of Blaxhall Church and at a quarter of a mile east-south-east and south-west of Stone Farm.

In the outlying patch of Boulder Clay west of Blaxhall Church there are three pits showing the sand beneath, one is on the north of Coal Pit Wood (a name suggesting that lumps of bituminous shale in the Boulder Clay have been regarded as evidence of coal to be found below), the others are marked on the map. A smaller outlier occurs between this and the railway.

The northern end of the spur of Boulder Clay extending towards Blackstock Wood is marked by a junction-pit showing the usual characters, and there is an outlier between this and Campsey Ash Church.

On the south-west of the Deer Park the Clay Pit marked on the map touches sand, and there is another pit about 200 yards southward of the church.

The extreme thinness of the Boulder Clay around Campsey Ash, and the inequalities of level of the surface on which it lies, cause the boundary to assume a very sinuous form, often without reference to existing surface-contour.

Sand is seen under the clay about 300 yards north of the railway-station. East of the Well House a pit penetrates to sand through clay, whilst on the north of the road at the same level, or perhaps somewhat higher, the sand forms the surface. A quarter of a mile to the north of the Well House are pits on either side of the road, the western showing 20, and the eastern 15, feet of Boulder Clay, over sand. A long narrow spur of the clay runs to the eastward.

About 200 yards northward along the road is seen a deposit of angular flints, worked in a shallow pit, and representing a rather exceptional state of the Boulder Clay, whilst the clay, here wanting as matrix, may be seen as a mixture of brown loam and dark sand in a brickyard a quarter of a mile to the westward. Boulder Clay of normal character is seen nearly all round these peculiar forms.

Good junction-sections occur on both sides of the high road half a mile west of Marlesford Station.

There is a small outlier of Boulder Clay west of Hacheston Church, and at the village a larger one, with a fine pit showing the junction with the gravelly sand below.

A similar pit may be seen about 300 yards south of Bloomville Hall. In the second hollow to the north, the base is exposed in an old pit, about 200 yards south-west of Parham Old Hall, and again at the Hall.

Thence the boundary descends to the river opposite Parham Wood, and the junction is seen in a pit by the side of the road. With the exception of a slight rise, at and opposite Broadwater, the bottom of the valley is occupied by Boulder Clay from here to near Framlingham.

Sand appears to have been reached below the clay in the abandoned brickyard south of Hill Farm, and the sections described on p. 33, show a temporary rise of the base south of Framlingham Station. In the town the very irregular surface of the sand produces two inliers of the sand and gives rise to unexpected complications in the draining and water-supply of the town, which latter depends (1876) upon private wells and upon a pump by the river-bank.

The low ground near the station is Boulder Clay, found by a well at the mill to be 25 feet thick.

The inliers of sand to the west have been mentioned on p. 33, and the base of the clay is seen in the nearer of the two, in a pit on its northern side.

West of the Mere the base rises towards the College, but is 40 feet down in the College well, and but little above the Mere where last seen at Little Lodge.

Valley of the Ore, down the Left Side.

At the back of Framlingham College is a shallow pit in sharp angular gravel and sand with chalk pebbles. As the Boulder Clay in the well close by is over 40 feet thick, this deposit may be regarded as a modification of the clay.

In the wood eastward of Little Lodge (to the north), and in the adjacent fields, there is seen gravel, which may be either an exceptional condition of the Boulder Clay (i.e., stones without clay matrix) or a protruding boss of the series beneath. The latter is again touched in the bottom of the pond in front of the Castle Brewery (through Boulder Clay) and is only 6 feet from the surface in front of the Crown and Anchor Hotel, whilst a well near by penetrated 40 feet, and another in Double Street 50 feet, of Boulder Clay.

The sections on the railway half a mile southward of the station, as described on p. 34, show similar irregularities on a smaller scale.

The base is next seen, on the left side of the valley, at the Rifle Butts, half a mile north-west of Broadwater, and thence runs southward with greater uniformity. Below Broadwater it produces a marked feature, and is seen in a pit south-east of Cole's Green Farm : there are two other such pits near Parham House and a fourth on the brook west of Parham Green. Opposite Parham House and east of it, are two more junction-pits of the usual type.

The railway-cutting north of Parham shows the base of the Boulder Clay on each side, and the sand is also touched in three large pits between the cutting and Parham Wood.

To the north-west of Parham Hall the base descends sharply, and is seen in an old pit on the east of the railway, and in the adjacent cutting the Boulder Clay occupies a trough eroded in the Red Crag sand. Southward of this are the pits described on p. 34, whence the base rises eastward, and then descends sharply towards the stream between Garden Cottage and Red Barn, a separated patch of clay lying on the slope.

Pits showing the junction of Boulder Clay and sand are found in the plantation north-west of Marlesford Hall, beyond the road east of the park, and on the road a quarter of a mile west of Little Glemham Manor House. The road-cutting west of Cotton's Barn shows the base of the clay, as does a pit a quarter of a mile to the south-east, and one at Cotton's Barn. Other similar sections are afforded by pits at 200 yards east of the Manor House, three on the south of the road between Glemham Hall and Marlesford Common, one half a mile west of Little Glemham Church, three between the Parsonage and the Church, and one a quarter of a mile east of the Church.

Valley of the Alde.

At the Keeper's Lodge, on the south-east of Great Wood, north of Little Glemham, a well showed 27 feet of Boulder Clay, whilst to the north of the wood the sand comes up as a narrow inlier. Opposite the mouth of the hollow which shows this, is another junction-pit, and there are three more in the neighbourhood of the Methodist Chapel, Stratford St. Andrew.

Outlying traces of the clay are seen covering sand in pits in the north-eastern part of Little Glemham Park.

At the Manor House, Stratford St. Andrew, is a pit showing the junction, and half a mile to the north-west is a similar pit.

Great Glemham Church stands on the edge of the clay, or just off it, and a large pit some 200 yards to the N.W. shows the junction of sand and clay. A similar section may be seen half a mile W.N.W. of the church, and another below the group of cottages a quarter of a mile to the N.W. The base of the clay is also exposed at the southern corner of High Grove, further N.W., and in two sand-pits between this and Glemham House.

The next noteworthy section is at White Barn, south of Sweffling, where, however, the junction is not seen, though there are pits in both series near each other. A quarter of a mile south-west of Sweffling Church is another junction-pit.

The base is next exposed in an old pit a quarter of a mile east of Sewers Cottage, N.W. of Sweffling, and it runs evenly thence to Cransford Hall, whence it descends gradually, and half a mile north-by-east of Cransford Church the junction is seen in a pit nearly at the bottom of the side-valley. On the north of this the Boulder Clay extends to the road opposite Bruisyard; and the base is exposed in a pit behind a farm opposite The Rookery. Thence it rises westward, and is met with in a large pit at Hucklin Hall.

Three-quarters of a mile north of this it is seen on the eastern side of the brook descending from Laxfield by Baddingham (in 50 N.E.). From the junction-pit south of Coldstone Hall, described on p. 34, the base runs, with great uniformity of level, to Bruisyard Hall, where it is again exposed.

At Rendham Grove Farm it begins to descend, reaching the alluvium of the Alde near Dobson's Farm. South of Potash Cottage (a name suggestive of woodlands and clay soil) the base again rises, and on three sides of Dodd's Wood there are pits showing the junction.

At and southward of the Pottery (by the river, west of Benhall), there are two more junction-pits, and others north and east of Benhall Street, one of the best lying to the north of the road to Benhall Lodge.

The sheet of clay eastward of Farnham is very thin, and is penetrated by large sand-pits, not only on the margin, but near the centre also.

Saxmundham Valley.

An outlying patch of clay occupies a somewhat lower level on the east; and northward of Farnham, the hollow of Benhall Park carries the base back, and nearly converts the clay on the south into an outlier.

Southward of Benhall Church is an inlier of the sands, with a pit, already described, at its eastern end (see p. 22).

North of the Rectory the base of the clay is shown in a pit, and in the railway-cutting to the east the junction with the sand is obscurely seen. It is better shown in a pit between the high road and the cutting, and in two pits, one on each side of the railway, westward of Bigsby's Corner, the western of which has been described on p. 22. Another junction-pit is seen near the railway some 650 yards south of Saxmundham Station.

The base is seen in many places on the western side of the valley above Saxmundham, viz., near the mill west of the station, 200 yards north-west of the Poor House; 400, and again 600, yards east of Spark's Barn; 200 yards south of Carlton Church; near the Keeper's Lodge; at the head of The Gull Stream; in three pits between the last and Carlton; at two points westward of Kelsale; and at Parkgate Farm.

On the east of the valley there are similar pits 400 yards north and 100 north-west of Kelsale Church and opposite Carlton Hall.

Northward of Sternfield the base is exposed in a series of pits from the point opposite Bigsby's Corner eastward up the lateral valley; on the eastern side of which there are four more similar pits between Poor House and Moor Farm. The base is shown in the road-cutting south of Sternfield, and in a pit in the outlier three quarters of a mile to the southward. Half a mile south of this two more junction-pits are seen in the westward spur of the clay, and five others near Snape Church. The most northern of the five is marked on the map as Sand and Clay Pit, two lie a quarter of a mile south-east of it, and one at the same distance to the south-west; from the last 200 yards towards Snape Church lies the fifth.

Valley of the Ore or Alde, Left Side.

The base of the clay is next seen south of Friston Hall, and again in the pit, described on p. 23, near Lichfield House. On the north of the hollow running up from Friston the base is exposed a quarter of a mile south-south-west of High House Farm, and in another pit half-way between this and the church. Two pits south of Laurel Cover lie on or close to the boundary, which here runs back so far east as almost to separate the south-eastward extension of the clay from the main mass of the deposit.

Turning westward again, for a little, the boundary passes through two pits on the footpath between Friston and Park House Farm, leaving another, on the road to Coldfair Green, to the left. Rounding the little hollow above Park House Farm, another junction-pit is passed, about 200 yards north-east of the farm-buildings.

Eastward of Hazlewood Hall, the pit at which has been noted on p. 34, a smaller section is seen on the junction at the next farm, and in a pit some 180 yards north of it. Thence the boundary becomes obscured by drifting sands, as the hillside curves round out of the Ore Valley, but two or three old pits, carried through the clay to the sand, show approximately the limits of the upper bed.

Valley of The Hundred River.

The base of the Boulder Clay continues obscure to a point half a mile south-west of Aldringham Church. Here the base is seen in a pit on the edge of the sand, whilst from Stone Cottage northward to the road is a line of four similar pits within a quarter of a mile.

A patch of Boulder Clay occurs in the lane half a mile E.N.E. of Billeaford Hall (E. of Friston), and there are traces of the clay on the intervening slope, so that the isolated exposure is probably a fragment of the main mass, and not a lenticular bed in the underlying sand.

The pit at the Hall has been described on p. 34, along with that to the west. There is another pit, reaching sand, between Billeaford Hall and Park House Farm. The narrow neck of Boulder Clay east of Friston has a junction-pit in it, and there are three more westward and northwestward of Knoddishall Church.

The junction is next seen by the side of the railway, and thence the boundary runs back to Knoddishall, passing one junction-pit on its way, and being exposed in another about 50 yards from the church. Another pit, a quarter of a mile east-north-east of the church, shows how closely the base of the clay follows the contour of the surface, running at a uniform level round the hollow eastward, and turning west to form a narrow spur north of Coldfair Green. This spur shows four junction-pits, two on the north, one at the western point, and one on the south. The base is also seen in pits south-west and north-west of Bedwells Farm. The sand-pit marked on the map half a mile to the north-east is worked through the edge of the clay, which thence skirts round to Leiston Church, 100 yards south-east of which is another junction-pit. North of the railway is an outlier with two pits carried into sand.

Half a mile north of the church, and again a quarter of a mile further, as well as at each western corner of the inlier of sand, the junction is exposed in pits.

South and south-east of Leiston Abbey is seen similar evidence of the course of the boundary, which passes close to Upper Abbey, and then, turning into the Minsmere Valley, forms a series of three short spurs, on the edges of which, and sometimes further in, are the usual junction-pits.

Minsmere Valley.

From above Theberton House, near which the boundary passes the section described on p. 35, it runs westward, with five junction-sections very close together about half a mile west-south-west of the House. Thence it runs in and out, according to the contour of the surface, past Theberton (where the base is seen a quarter of a mile west of the high road, and again about 200 yards west of Theberton Hall), and far up the side-valley on the north. The section showing its position half a mile west-north-west of the Hall has already been described (p. 35), and the junction is again seen at the sand-pit, marked on the map, a quarter of a mile beyond.

The outliers north-east of Theberton Church and Hall contain pits of like character.

The long spur extending to Middleton attains a low level near the church, and a mass of Boulder Clay, in the lane skirting the alluvium north of the Plough Inn, seems to be an outlier at a still lower level.

On the high 'road three quarters of a mile west of the Plough the base is again seen in a pit, also at three points southward, and two north-eastward, of Fordley Hall.

Lastly, a side-hollow, running up from the valley below Yoxford, enters this district on the east of and close to the railway and shows the base of the clay just within Sheet 50 S.E.

Valley of the Waveney.

Two small inlying patches of sand protrude through the Boulder Clay east of Bedfield, and give rise to a stream flowing by Tannington to join the Waveney, by a circuitous course, near Hoxne.

CHAPTER VII.—POST GLACIAL BEDS.

The river-gravels and alluvial deposits of the district hardly require notice beyond the delineation of their boundaries on the map. They are of course formed from the beds of the higher land and consist of the same material as these, more finely divided and re-arranged.

RIVER GRAVELS.

But small tracts of these gravels occur, and none of the long strips common along some valleys have been found. These patches of low-lying gravel occur in the valley of the Deben from Wickham Market to Woodbridge, and in the valley of the Ore from Snape Bridge to Iken. A very small patch has also been mapped about a mile N.E. of Butley Church.

By the railway east of Woodbridge Church a pit, 13 feet deep, reached through the gravel to the London Clay, and some large bones were found about half way down, but they were not identified.

ALLUVIUM.

Strips of marshland occur along the streams, and they widen out to broad marshes by the coast, along the greater part of which indeed the land is of this kind, and below high-water level.

The only sections noted are at Framlingham, where "In April 1823 some labourers who were raising gravel on the Little Lodge lands, through which the river runs, dug up at *ten feet* below the surface, and level with the bed of it, two elephant's tusks."* The overlying 10 feet may have been stony rainwash or earth slipped from the steep banks.

The alluvium of Framlingham Mere, near the Castle, gave the following section for which we are indebted to Mr. LANE :—

Made earth, &c.
Blue silty clay
Sand
Pebbly sand with antlers of deer, abundance of estuarine (?) shells, &c. ⎱ 7 feet.

Possibly the shells are *Uniones*, mistaken for *Myæ:* unfortunately they were not preserved.

Southward from the end of the Minsmere cliff, clay and peaty earth (with roots, etc.) are seen on the fore-shore beneath the beach. These are the seaward prolongation of the alluvium, the so-called submerged forest that is often seen in this position.

---

* R. GREEN's "History, &c. of Framlingham and Saxsted." 8vo. *Lond.*, 1834, p. 11.

Coast Deposits.

Shingle.

From Orfordness, at the southern edge of our district, to Aldborough, is a continuous sheet of shingle, piled up by the sea in the course of time. It is very narrow on the north, but broader on the south, whence it extends south-westward, dividing the river from the sea, to Orford Haven, as described in the Memoir on the map to the south,* and one cannot do better than quote the description of the whole given by Mr. J. B. REDMAN, in his paper on the East Coast.†

"This extraordinary mole of shingle is 10 miles [rather more now] in length, parallel to the shore . . . . . . The general direction of Orford Beach to the south of the Ness is S.W., and northward, Aldborough beach bears N.N.E. Orfordness is formed of a series of curved concentric 'fulls,' sweeping round and forming a projecting point in advance of the general coast line. . . . . The triangular projection thus formed encloses salt marshes next the rivers. . . . . Off the Marsh House, midway between the High Light and North Weir Point [as it then was], and towards the 'Narrows,' there is a local twist, or change in direction, of every successive ridge of shingle. This has produced a series of hummocks east of Orford Water, opposite Orford Quay and Castle. . . . ."

"Very large pebbles are found on the summits of these ancient 'fulls,' which become lower in-shore, and which, curving round towards the broad water opposite the Castle, denote the successive terminations of the beach at early periods; each successive horn shutting in a deep valley between it and the beach upon which it has formed; these terminations still remain and are distinctly traceable." . . . .

"Between the two Lights . . . . there is a modern projection of the Ness, of a brighter, yellower appearance than the rest of the mass. This accumulation, which is the growth of the last twenty years [written in 1863 or earlier], is entirely local, and has been made in a series of convex curves overlapping each other, and forming a point of from 200 yards to 300 yards in extent."

"The general formation of Orfordness is like Dungeness, and like it has deep water close up to the point; and again, similar to Langley Point, the ancient fulls are intersected at an angle by modern formations. Very large flint boulders are found on the summit of the Ness behind the High Light. The local increase at the point has been accompanied . . . by a corresponding decrease . . . . off the Low Light and towards Aldborough."

. . . "In three centuries the shingle appears to have travelled south-westward 5 miles, giving an average annual leeward progression of 30 yards. This rate must have been even greater over certain periods, as for some years past [1863?], the outlet has been stationary, or rather it has receded northwards, and forced its way through a weak part of the beach, or possibly through a partial breach caused by the sea."

With the author's consent, and by the kindness of the Council of the Institution of Civil Engineers, a reproduction is given of MR. REDMAN'S views of the beach from Orfordness (Figs. 10, 11, of his paper). These woodcuts cannot fail to assist the reader in appreciating the character of this remarkable shore-accumulation.

---

* The Geology of the Country around Ipswich, &c., pp. 90, 99. (1885.)
† *Proc. Inst. Civ. Eng.*, vol. xxiii., pp. 200, 203. (1865.)

FIGS. 8, 9.

Views from the Top of Orfordness High Lighthouse. (REDMAN, 1863.)

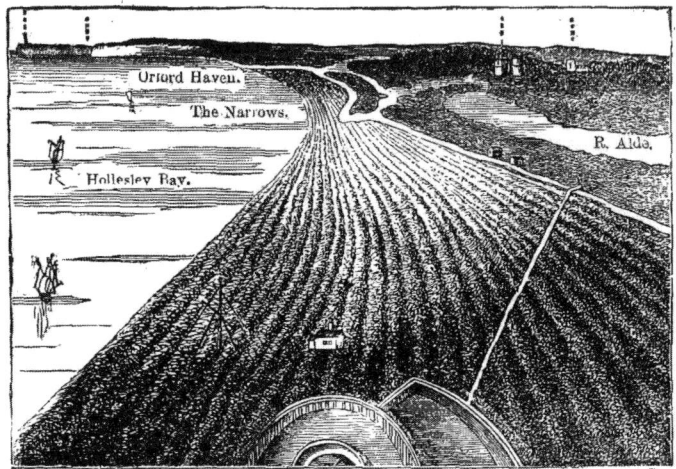

Looking S.W. Showing a local twist in the shingle-fulls E. of Orford Waters, E. of the Old Haven. The direction of the fulls, along Hollesley Bay, is E. by N. to W. by S.

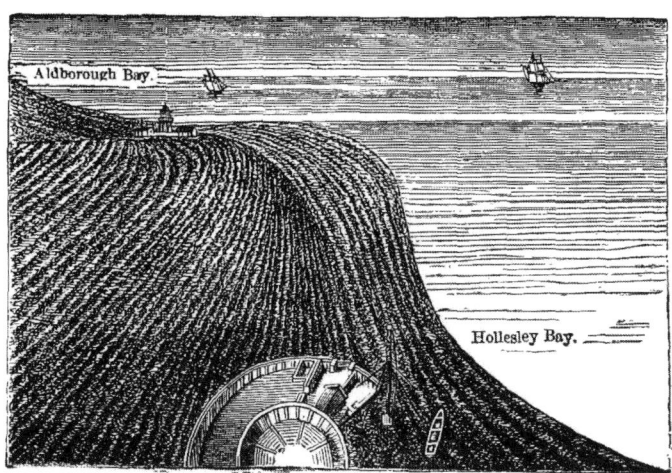

Looking N.E. by E. Showing the modern spit, of 18 years growth, between Aldborough Bay and Hollesley Bay, 200 to 300 yards in breadth. The direction of the old fulls is W. to E., and, near the High Lighthouse, W. by S. to E. by N.

From Aldborough to Thorpe is a narrow strip of shingle, through which, near the middle, the Hundred River passes. Of the northern end of this, Thorpe Ness, Mr. REDMAN remarks that "it appears to have grown out to the extent of from 30 yards to 40 yards during the last few years." [1863?]*

From Sizewell to the cliff north of Minsmere Level is another strip of shingle, with some blown sand.

Blown Sand.

There is a short patch of sand, blown up from the foreshore by the wind, at Thorpe; and MR. REDMAN has described the marsh of Minsmere Level as "fringed by a wide belt of low drifted sand-hillocks a furlong broad, then two series of shingle, old, and modern, south of the sluice . . . . the beach and sandhills decrease northward . . . . . where the modern beach is retreating upon the base formed by the early sand 'dunes'."† This coast is well calculated to supply sand, and to the north the pebbly gravel supplies shingle. W. W. and W. H. D.

---

* *Proc. Inst. Civ. Eng.*, vol. xxiii., p. 208.
† *Proc. Inst. Civ. Eng.*, vol. xxiii., p. 204.

APPENDIX.—WELL-SECTIONS.

By W. Whitaker and W. H. Dalton.

Aspal.

Rev. W. B. Clarke, *Trans. Geol. Soc.*, ser. 2, vol. v., p. 377.

Water within 40 feet of the surface.

[Drift] { Diluvial [Boulder] Clay, 70 feet.
{ Sand with water, 10 feet.

Boulge Hall.

Ipswich Journal, Aug. 3, 1872, p. 5.

Bad water, containing organic phosphates, shut out by pipes in the upper part. Large supply of good water got from the Chalk.

To Chalk, 160 feet.
In ,, 90 ,,
―――
250 ,,

Brandeston Hall.

A. Legrand. *Trans. Soc. Engineers* for 1877, p. 141.

? Old well, 16 feet.
Clay, about 20 ,,
―――
To Chalk 36 ,,

Easton Park. About a quarter of a mile N.E. of the house. Information from Mr. D. Smith (Letter, Dec. 1875).

		Feet.
Old well [? 1845] gave good supply at about		65
Bored in 1873? { Very coarse stony sand	about	30
Firm sand	,,	5
	To Chalk ,,	100
Chalk		400 to 500

The old well was probably mostly in Boulder Clay.

From a letter from Mr. Smith to Mr. S. V. Wood, Junr., dated April 1874, the following further particulars are taken. The sinking went through a considerable thickness of clay into sand, with a good deal of iron. Bored in 1873 in sandy material, with slight traces of Crag. Total depth, 550 feet.

Framlingham. Albert College, 1875.

About 150 feet above the level of the sea.

From information from workmen on the spot.

Old well 68 feet. Bad water soaks through brickwork at 40 feet [indicating the base of the Boulder Clay]. Bored below this. Water rose at first to 50 feet from surface; in 1879 it had fallen to 58 feet.

	Thickness.	Depth.
Old well { Boulder Clay - - - 40 ? } { Sand, &c. - - - 28 ? }	68	68
Grey, pebbly, very sharp sand, full of impure water	27	95
Chalk - - - - - -	135	230

FRAMLINGHAM.

1. Near Crown-and-Anchor Inn. 2. In Double Street. 3. Near Railway Station.

Information from MR. J. BARKER.

	1.	2.	3.
Boulder Clay	40	50	25 feet.
	to sand.	to gravel.	

FRAMLINGHAM. An old well, without precise locality, given by REV. W. B. CLARKE. *Trans. Geol. Soc.*, ser. 2, vol. v., p. 377.

Diluvial [Boulder] Clay, full of septaria and [derived] fossils	15 feet.
Sand	25 ,,

FRAMSDEN. Brickyard N. of Helmingham Park (for Lord Tollemache). Sunk and communicated by MESSRS. BENNETT, of Ipswich.

Water rises to within 23 feet of the surface; supply good.

		THICKNESS.	DEPTH.
Mixed soil		7	7
[Drift 104 feet.]	Gravel	1	8
	Boulder Clay	21	29
	Chalk stone or marl	1½	30½
	Light [-coloured] sand	½	31
	Boulder Clay	71	102
	Brown loaming clay	2½	104½
	Light [-coloured] running or blowing sand	6	110½
	Black flints	½	111
Chalk		99	210

GREAT GLEMHAM.

REV. W. B. CLARKE *Trans. Geol. Soc.*, ser. 2, vol. v., p. 377.

Diluvial [Boulder] Clay	20 feet.
Sand	20 ,,

HASKETON. Cottage at the S. end of Blunt's Wood, on the edge of the map, ¼ mile N. of Lechford Hall (in 48, N.E.).

S. V. WOOD, JUNR. *Quart. Journ. Geol. Soc.*, vol. xxxiii., pp. 106, 107. (1877).

			THICKNESS.	DEPTH.
Glacial Drift.	Boulder Clay		6	6
	Sand and gravel.	Coarse gravel	3	9
		Fine gravel, with fragments of marine shells, passing into the bed below	3	12
		Buff sand	45	57
Red Crag	Crag with shells		6	63
	Crag with water		4	67

HELMINGHAM HALL.

REV. W. B. CLARKE, *Trans. Geol. Soc.* ser. 2, vol. v. p. 377.

	THICKNESS.	DEPTH.
Mould	1	1
Clay with [derivative] shells [Boulder Clay]	56	57
Chalk [? chalky Boulder Clay or gravel]	30	87
Dark sand	5	92

LEISTON. Ironworks.
Information from MR. GARRETT, (from memory).
About 60 feet above the sea.
Water level 34 feet down.

	THICKNESS.	DEPTH.
Brickearth, clay, and sand	45	45
Blowing sand (yields 8–10 gals. a minute)	5	50
Brickearth, clay, and moulding sand	80	130
Blue shelly Crag (yields 80 gals. a minute)	60	190

[Borehole now filled with shelly sand, but it must have very nearly reached the Tertiary clay, as the estimated position of the Chalk surface is only 229 feet down.—W. H. D.]

MELTON. Brewery (Mallet's). 1874.
Sunk and communicated by MESSRS. BENNETT of Ipswich and information got by W.H.D.
About 28 feet above the level of the sea.
Good supply of water to 21 feet from the surface.

Depth of old well (the rest bored)	25 feet.
Sand, greenish and clayey towards the bottom	35 ,,
To Chalk	60 ,,
Chalk	151 ,,
	211 ,,

MONEWDEN. Mr. Haggar's (plumber). 1876.
Sunk and communicated by MESSRS. BENNETT, of Ipswich.
Shaft throughout.
Good supply, 5½ feet from the bottom.

			THICKNESS.	DEPTH.
[Glacial Drift.]	[Boulder Clay].	Light-coloured clay and stones	10	10
		Large stones	1	11
		Blue clay and chalk stones	19	30
	[Glacial Gravel, &c.]	Blue loam	3	33
		Coarse gravel and stones	12	45
		Clay and marl	5	50
		Sand	¼	50¼
[?Chalk.]		Chalk	1	51¼
		Hard craig mixture [? hardened brecciated chalk]	2	53¼
		Chalk	7¾	61

ORFORD. Lantern Marshes, N.E. of the Town. 1875.
About 6 feet above Ordnance Datum.
Sunk and communicated by MESSRS. BENNETT, of Ipswich.
Water overflows.

		THICKNESS.	DEPTH.
Depth of old bore		—	100
[London Clay].	Blue clay	65	165
	Dark sand	2	167
	Brownish clay	10	177
	Hard rock	1	178
[Reading Beds?]	Dark loam	3	181
	Brownish clay	20	201
	White and red sand	4	205
	Clay	18	223
	Flints	1	224
Chalk		96	320

ORFORD. Lantern Marshes, No. 2. 1876.

Sunk and communicated by MESSRS. BENNETT, of Ipswich.

Supply ample but very brackish, rose 2 feet above surface; in August 1878, as copious less brackish; in June 1879, still better but had to be shut off during wet season.

		THICKNESS.	DEPTH.
[Alluvium.]	Ooze	24	24
[? Red Crag.]	White sand and shell	7	31
[London Clay, about 129 feet.]	Blue clay	97	128
	Rock	1	129
	Dark brown clay	30	159
	Rock	1¼	160¼
	White sand	6	166¼
	Clay	14	180¼
[Reading Beds, about 50 feet.]	White sand	5¼	185½
	Red and green running sand [Specimen of red and grey mottled sand, from 186 feet]	8½	194
	Blue clay	16	210
Chalk		70	280

ORFORD. Marshes, (Lord Rendlesham's), 1878, N.B., *not* well No. 3 of the map, the account of which will be found in the Memoir on 48 N. p. 121.

About 6 feet above Ordnance datum.

Sunk and communicated by MESSRS. BENNETT.

		THICKNESS.	DEPTH.
[Alluvium.]	Soft clay	30	30
[Red Crag.]	White crag	2	32
[London Clay, 108 ft.]	Brown clay	4	36
	London clay	23	59
	Rock [septaria]	1	60
	London clay	25¼	85¼
	Rock [septaria]	¾	86
	London Clay	12	98
	Rock, very hard [septaria]	1	99
	Brown clay	37	136
	Soft loam (light colour)	4	140
[Reading Beds.]	Very fine sand (nearly white)	12	152

After getting down some 160 feet this well was abandoned owing to failure in one of the tubes.

PARHAM. No precise locality.

REV. W. B. CLARKE, *Trans. Geol. Soc.*, ser. 2, vol. v., p. 377.

Diluvial [Boulder] clay, 25 feet.
Sand [? partly Crag], 30 „

SAXMUNDHAM. Messrs. Waller's Brewery. 1876.

Surface level 45 feet above the sea.

From information on the spot.—W. H. D.

Water rises to 12 feet from surface.

		THICKNESS.	DEPTH.
Red Crag.	Sand	80	80
	Red shelly Crag	6	86
	Bluish-black Crag	19	105
Reading Beds.	Stiff blue clay	19	124
	Stiff green clay	1½	125½
	Stiff brown clay	½	126
Chalk		53	179

The account furnished by Messrs. Bennett, of Ipswich, varies but slightly from the above, as follows:—

	Thickness.	Depth.
Old well	—	16
Red running sand	59	75
Crag	30	105
Clay	21	126
Chalk	53	179

Messrs. Bennett showed me specimens of brown sandy clay from 105 and 107 feet deep, and of green sandy clay and green-coated flints from just above the Chalk. W. W.

Wickham Market.

Rev. W. B. Clarke, *Trans. Geol. Soc.*, ser. 2, vol. v., p. 377.

Blue Clay	50 feet.
Sand	
Clay	
Sands and gravel	110 ,,
Chalk	

Witnesham.

Rev. W. B. Clarke, *Trans. Geol. Soc.*, ser. 2, vol. v., p. 379.

Clay	17 feet.
Gravel?	46 ,,
To Chalk	63 ,,

Woodbridge. Mr. Hayward's Steam Flour Mill, close to the Gasworks. 1874.

Sunk and communicated by Messrs. Bennett, of Ipswich.

Good supply; about 40 gallons a minute, rising to within 8 feet of the surface.

		Thickness.	Depth.
Depth of well (the rest bored)		—	15
[Reading Beds]	Sand	11	26
	Blue clay	4	30
	Sand	11	41
	Blue clay	2	43
	Sand	1	44
	Mixed clay	4	48
	Small flints	$\frac{1}{4}$	$48\frac{1}{4}$
Chalk		$71\frac{3}{4}$	120

Woodbridge. Mr. Gall's, The Thoroughfare, 1876.

Sunk and communicated by Messrs. Bennett, of Ipswich.

Water rises to within 25 feet of the surface. Level not affected by pumping 200 gallons an hour.

		Thickness.	Depth.
Depth of old well (the rest bored)		—	30
[Reading Beds]	Sand	14	44
	Clay	2	46
	Loamy sand	13	59
	Clay	3	62
	Flints	1	63
Chalk		79	142

WELL-SECTIONS.

WOODBRIDGE. Messrs. Hart and Wrinch, maltsters, by the river. 1876.
Sunk and communicated by MESSRS. BENNETT, of Ipswich.
Water rises to 21¼ feet below the surface. Good supply, 70 gallons a minute.

		THICKNESS.	DEPTH.
Depth of old well (the rest bored)		—	23½
[Reading Beds]	Fine sand	14	37½
	Mottled clay	6	43½
	Loamy sand	14	57½
	Dark clay	3	60½
	Brown flints	1	61½
Chalk		78½	140

WOODBRIDGE. Farm ¾ mile N. of church.
Communicated by MR. S. V. WOOD, JUNR. 1881.
Water got at the bottom.

	THICKNESS.	DEPTH.
Humus and sandy brickearth	2	2
White Boulder Clay, very chalky	4	6
Coarse sand, full of chalk grains, and small lumps of chalk, mostly very hard	12	18
Sandy gravel, with but few large stones and many shell-fragments and small lumps of hard chalk, the last getting scarcer downwards, and ceasing at the base	14	32
Sandy gravel with few (and those small) stones, no shell-fragments or chalk	21	53
Coarse sand with white grains (? chalk)	12	65
Yellow gravel, more stones, but none large	5	70
More sandy gravel	6	76
Dark yellow sand without gravel*	4	80
The same, darker and with some comminuted Crag in it*	3	83

* These two beds were regarded by MR. WOOD as of Lower Glacial age, the Crag being remanié.

WOOODBRIDGE. Mr. Combe's Malting on the top of the hill, near the Sun Inn. 1885.
Made and communicated by MESSRS. BENNETT.
Water rises to within 11 feet of the surface. A good supply.

		THICKNESS.	DEPTH.
Shaft, partly in blowing sand (the rest bored)		—	13½
[? Crag]. Very fine running sand		9	22½
Blue [? London] Clay		2½	25
[Reading Beds].	Running sand	14	39
	Mottled clay	2½	41½
	Running sand	1½	43
	Light-coloured loam	2½	45½
	Red mottled clay	5½	51
	Dark green clay with brown flints	1	52
Chalk, very white and clean		88	140

This section, which was received after the MS. of the Memoir had been sent in, seems to bring the Reading Beds very near to the surface. Possibly the running sand that occurs 25 feet down may represent the sandy basement-bed of the London Clay; but on the other hand, the Reading Beds should perhaps be carried higher. W. W.

WOODBRIDGE. Notes on Well-sections in MR. CLARKE'S Paper. *Trans. Geol. Soc.*, ser. 2, vol. v., pp. 382, 383.

The section, given as that of a well at the Post Office, must be wrong, making the depth to the Chalk 510 feet! Nor can the site be identified: it is not at the present Post Office.

The "Barrack Ground" of another section is also non-existent, and a third, "at the entrance of the town," has too vague a site assigned.

Since the publication of the Memoir on the country to the south* two Suffolk well-sections in Sheet 48, N.E., have come to hand, which it is thought best to print here as a supplement to that Memoir.   W. W.

FELIXSTOW.   Messrs. Bugg and Colley. At the foot of the cliff between Bent Hill and the Bath Hotel. 1885.

Sunk and communicated by MESSRS. BENNETT, of Ipswich.

Shaft, 28 feet, the rest bored.

A good supply of good water.

			THICKNESS.		DEPTH.	
			FT.	IN.	FT.	IN.
[London Clay, 64½ feet.]		Clay with septaria at four levels	56	7	56	7
	[Basement Bed.]	Running sand	2	4	58	11
		Loam and flints [? pebbles]	5	0	63	11
		Rock	0	6	64	5
[Reading Beds, 43½ feet.]		Mottled clay	7	0	71	5
		Light sandy loam	8	0	79	5
		Mottled clay	2	0	81	5
		Light sandy loam	3	0	84	5
		Mottled clay	6	0	90	5
		Light sandy loam	7	0	97	5
		Slate-coloured clay	3	6	100	11
		Mottled clay	3	0	103	11
		Grey clay	1	0	104	11
		Mottled clay	1	6	106	5
		Flints	?1	7	108	0
Chalk			69	0	177	0

IPSWICH.   Messrs. Paul's, Smart's Wharf, St. Mary Key. 1885?

Sunk and communicated by MESSRS. BENNETT.

Shaft, 11 feet, the rest bored.

Salt water came in through the upper soft part of the Chalk, so that tubes had to be driven to a depth of 117 feet. This effectually excluded the salt water, and a good supply of fresh water is got, rising to within about 7 feet from the surface.

		THICKNESS.	DEPTH.
Mixed soil		11	11
[Drift, 29½ feet]	Coarse gravel [? River gravel]	15	26
	Boulder Clay	12	38
	Gravel and sand	2½	40½
Chalk, the upper part very soft		148½	189

This well is within 200 yards of that at the New Mill, St. Peter's Quay (described in the Memoir on the Geology of Ipswich, p. 118); but the sections of the two differ very much.

---

* The Geology of the Country around Ipswich.   (1885.)

The following section was accidentally left until after sending to press, and is therefore given here, out of its proper place:—

KELSALE. Red House Farm, [? halfway between Carlton and Rendham.]

Communicated by Mr. F. H. VERTUE, of Southwold.

	THICKNESS.	DEPTH.
Mould and bright clay, with water of fair quality	12	12
[Glacial Drift.] Gravel	3	15
[Glacial Drift.] Bluish [Boulder] Clay, with chalk-stones	90	105
[Glacial Drift.] Bright sand - about	10	115
[Crag.] Black muddy sand, with a disagreeable smell, and a plentiful supply of clean water	?	?

The classification of the sands is somewhat doubtful. Possibly both may be Drift; but, on the other hand, both may be Crag. Along the neighbouring outcrops the Boulder Clay is next underlain by Glacial sand; but that clay extends to such a depth in the well that it may have cut through to Crag.　　　　　　　　　　　　　　　　　W. W.

INDEX.

* is prefixed to names of places out of the district.

Aldborough, or Aldeburgh, 1, 2, 8–10, 23, 24, 26, 29, 47, 49.
Aldborough Bay, 48.
Alde, River and Valley (see also Ore), 1, 2, 20–23, 28, 29, 33, 34, 37, 41–44, 48.
Aldringham, 24, 26, 29, 35, 44.
Alluvium, 46.
Alteration (decalcification) of Red Crag, 12, 14, 15, 18, 19, 21, 24.
*Antwerp, 5.
Aragonite-shells, 9.
Ash Bocking, 2.
Ashfield, 3, 38, 39.
Aspal, 50.

*Baddingham, 43.
Barker, J., 51.
Base of Tertiaries, plane of, 3.
*Bawdsey, 48.
Bedfield, 45.
Bell, A. 6, 17.
Benhall, 22, 43.
Bennett, Messrs., 51–56.
Blaxhall, 28, 29, 34, 41.
Blown sand, 49.
Boulder Clay, 27, 36.
Boulge, 50.
*Boyton, 9.
Brandeston, 4, 28, 32, 38, 39, 50.
Bredfield, 1, 31, 37.
Brickyards, 4, 10, 15, 18, 21, 23, 26, 28, 29, 32, 33, 35, 38, 39, 41.
Bromeswell, 40.
Bryozoan Crag, 5.
Bure Valley Beds, 27.
Burgh, 4, 36.
Butley, 13, 14, 17, 18, 25, 46.
Butley River and Valley, 2, 4, 9, 10, 17–19, 28, 33, 40.
Bruisyard, 34, 43.

Calcite-shells, 9.
Campsey Ash, 4, 17, 33, 39, 41.
Carlton, 2, 43, 57.
Cavities, tubular, in Coralline Crag, 7.
Chalk, 3.
Charlesworth, E., 7.
Charsfield, 4, 37.
Charsfield Valley, 32.
Chillesford, 2, 18, 19, 26, 40.
Chillesford Clay, 25.
Clarke, Rev. W. B., 50, 51, 53, 54, 56.
Clay-lump (building), 36.
Clopton, 14, 31, 36.
Coast deposits and changes, 47.
Colour of Red Crag, 12, 20.
Concretionary structure in Red Crag, 12, 13.
Contorted Drift, 28.
Coprolite-bed, see Nodule-bed.
Coralline Crag, 5.

Crag, 5.
Cransford, 43.
Cretaceous Beds, 3.
Cretingham, 32, 38, 39.
Crisp, Dr. 19.
Cromer Till, 28.
Crawfoot, W. M., 25.
Culpho, 31, 36.

Dallinghoo, 37.
Dalton, W. H., 9.
Debach, 1.
Deben, River and Valley, 1, 2, 3, 4, 14–17, 28, 31, 32, 37–40, 46.
Debenham, 1, 32, 39.
Decalcification, see Alteration.
Dennington, 2.
Depth of Coralline Crag Sea, 8, 9.
Diestian, 5.
Dowson, E. T., 25.

Earl Soham, 1, 4, 16, 28, 32, 38, 39.
East Anglia, 9.
Eastern Counties, 8.
Easton, 3, 17, 32, 38, 39, 50.
Eyke, 17, 28, 32, 40.

Farnham, 2, 22, 43.
Faults, 23, 49.
*Felixstow, 56.
Filling in of whorls of Shells, 9–11.
Finn, River and Valley, 1, 4, 14, 15, 30, 31, 36, 37.
Fisher, Rev. O., 21.
Fitch, R., 7.
Forbes, Prof. E., 7.
Fossils of the Coralline Crag, 5.
Framlingham, 1–4, 21, 33, 41, 42, 46, 50, 51.
Framsden, 3, 32, 38, 51.
Friston, 23, 29, 34, 43, 44.
Frome, River = Ore.

*Gedgrave, 8–10.
Glacial Drift, 27.
Glacial Sand and Gravel, 30.
Gleme, River = Alde.
Glemhams, The, 2. See also Great and Little.
Godwin-Austen, R. A. C., 8.
Goodchild, J. G., 11.
Gravels, 30, 46.
Great Bealings, 12, 14, 15, 37.
Great Glemham, 4, 22, 42, 51.
Green, R., 46.
Grundisburgh, 1, 4, 14, 31, 36.

Hacheston, 32, 33, 38, 39, 41.
*Halesworth, 25.
Hanley, S., 7.
Harmer, F. W., 9, 12, 15.
Hasketon, 15, 28, 31, 36, 37, 51.
Heights of ground, 2.

Helmingham, 1, 38, 51.
High Suffolk, 2.
*Hollesley, 2.
*Hollesley Bay, 48.
Hoo, 38.
*Hoxne, 1, 45.
Hundred River and Valley, 2, 24, 34, 44, 49.

Iken, 2, 8, 10, 21, 26, 28, 46.
Inclined plane at base of Tertiaries, 3.
*Ipswich, 56.
Ironstone, see Limonite.

Jeffreys, Dr. J. G., 6, 7.

Kelsale, 1, 57.
Kendall, P. F., 9.
Kenton, 1, 2.
Kettleburgh, 32, 39.
Knoddishall, 2, 44.

Lane, —., 46.
*Laxfield, 43.
Legrand, A., 50.
Leiston, 1, 4, 24, 35, 44, 52.
Letheringham, 16, 22, 37, 38.
Limonite in Red Crag, 12, 14, 16, 22, 24.
Literature of the Coralline Crag, 7.
Little Glemham, 2, 22, 42.
London Clay, 4.
Low Suffolk, 2.

Marlesford, 1, 2, 21, 22, 33, 41, 42.
Melton, 3, 4, 15, 16, 31, 52.
Middleton, 24, 25, 35, 44.
Minsmere Level, River, and Valley, 2, 24, 25, 27, 35, 44–46, 49.
Monewden, 4, 16, 38, 52.
Monk Soham, 32.

Nodule-bed of Coralline Crag, 5, 9.
    "   "   Red Crag, 5, 9, 17, 20, 24.
North Sea, 1.

Ogilvie, C. P., 11.
Ore, River and Valley, 1, 2, 10, 19–23, 26, 28, 29, 33, 34, 37, 41–44, 46, 48.
Orford, 1, 2, 4, 7, 8, 10, 19, 20, 26, 47, 48, 52, 53.
*Orford Haven, 47, 48.
Orfordness, 47, 48.
*Orwell, River, 14.
Oscillations of Level, 8, 23, 25.
Otley, 1, 4, 14, 31, 36.

Parham, 22, 34, 41, 42, 53.
Pettaugh, 32, 39.
Pettistree, 4, 16, 28, 37.
Phosphate-bed, see Nodule-bed.
Physical features, 2.
Pipes in Coralline Crag, 7, 11.
Plane of base of Tertiaries, 3.
Post Glacial Beds, 46.
Potford Brook and Valley, 16, 37, 38.
Pottery, 43.

Prestwich, Prof. J., 5, 6, 8–10, 13, 18–21, 23, 25, 27.
Puddled Chalk, 3.

*Ramsholt, 7.
Reading Beds, 3, 4.
Re-constructed Chalk, 4, 29.
Red Crag, 12.
Redman, J. B., 47–49.
Rendham, 4, 43, 57.
Rendlesham, 1, 2, 4, 17, 33, 39, 40.
Reyce's Breviary, 2.
Rivers, 2.
River Gravels, 46.

Saxmundham, 1–4, 23, 43, 53.
Saxmundham Valley, 22, 23, 43.
Saxtead, 1.
Shingle, 47.
Sizewell, 2, 11, 24, 26, 30, 49.
Smith, D., 50.
Snape, 1, 4, 21–23, 29, 43, 46.
Soham Valley, 32, 39.
Sternfield, 23, 43.
Stopes, H., 11.
Stratford St. Andrew, 22, 42.
Submerged Forest, 46.
Sudbourn, 2, 7, 8, 10, 19–21, 26.
*Sutton, 5, 8, 9.
Swallow-hole, in Coralline Crag, 24.
Sweffling, 22, 42, 43.

Tannington, 1, 2, 45.
*Tattingstone, 8, 9.
Taylor, Dr. J. E., 17.
Tertiaries, inclined plane at base of, 3.
Theberton, 24, 25, 35, 44.
Thorpe, 11, 24, 49.
Tuddenham, 1, 14, 30, 36.
Tunstall, 21, 26, 28, 40, 41.

Ufford, 4, 16, 31, 37.

Van den Broeck, E., 5, 14
Vertue, F. H., 57.

*Walton (Naze), 14, 25.
Wantesden, 25, 40.
Water-supply, 38, 41, 50–56.
*Waveney, River, 1, 45.
Weathering of Boulder Clay, 35, 38.
Well-sections, 50.
Westleton Beds, 27.
Whale, remains of, 19.
Whitaker, W., 1, 2, 4–9, 24, 25, 27, 30, 34, 35, 46, 47, 49–56.
White Crag, 5.
Wickham Market, 1, 3, 17, 28, 31, 37, 46, 54.
Winston, 3, 16, 32, 39.
Witnesham, 1, 4, 14, 36, 54.
Wood, S. V., 5–9.
Wood, S. V., Jun., 9, 12, 14, 15, 17, 21, 24, 25, 27–30, 40, 50, 51, 55.
Woodbridge, 1, 4, 15, 28, 31, 46, 54, 56.
Woodward, S. P., 10.

*Yoxford, 45.

LONDON: Printed by EYRE and SPOTTISWOODE,
Printers to the Queen's most Excellent Majesty.
For Her Majesty's Stationery Office.
[10025.—500.—12/86.]

GENERAL MEMOIRS OF THE GEOLOGICAL SURVEY—continued.

The WEALD (PARTS of the COUNTIES of KENT, SURREY, SUSSEX, and HANTS). By W. TOPLEY. 17s. 6d.
The TRIASSIC and PERMIAN ROCKS of the MIDLAND COUNTIES of ENGLAND. By E. HULL. 5s.
The FENLAND. By S. B. J. SKERTCHLY. 36s. 6d.
The MANUFACTURE of GUN FLINTS. By S. B. J. SKERTCHLY. 16s.
The SUPERFICIAL DEPOSITS of SOUTH-WEST LANCASHIRE. By C. E. DE RANCE. 10s. 6d.
The CARBONIFEROUS LIMESTONE, YOREDALE ROCKS and MILLSTONE GRIT of N. DERBYSHIRE. By A. H. GREEN, DR. C. LE NEVE FOSTER, and J. R. DAKYNS. (2nd Ed. in preparation.)
The BURNLEY COAL FIELD. By E. HULL, J. R. DAKYNS, R. H. TIDDEMAN, J. C. WARD, W. GUNN, and C. E. DE RANCE. 12s.
The YORKSHIRE COALFIELD. By A. H. GREEN, J. R. DAKYNS, J. C. WARD, C. FOX-STRANGWAYS, W. H. DALTON, R. RUSSELL, and T. V. HOLMES. 42s.
The EAST SOMERSET and BRISTOL COALFIELDS. By H. B. WOODWARD. 18s.
The SOUTH STAFFORDSHIRE COAL-FIELD. By J. B. JUKES. (3rd Edit.) (*Out of print*.) 3s. 6d.
The WARWICKSHIRE COAL-FIELD. By H. H. HOWELL. 1s. 6d.
The LEICESTERSHIRE COAL-FIELD. By EDWARD HULL. 3s.
ERUPTIVE ROCKS of BRENT TOR. By F. RUTLEY. 15s. 6d.
FELSITIC LAVAS of ENGLAND and WALES. By F. RUTLEY. 9d.
HOLDERNESS. By C. REID. 4s.
BRITISH ORGANIC REMAINS. DECADES I. to XIII. with 10 Plates each. Price 4s. 6d. each 4to; 2s. 6d. each 8vo.
MONOGRAPH I. On the Genus PTERYGOTUS. By T. H. HUXLEY, and J. W. SALTER. 7s.
MONOGRAPH II. On the Structure of the BELEMNITIDÆ. By T. H. HUXLEY. 2s. 6d.
MONOGRAPH III. On the CROCODILIAN REMAINS found in the ELGIN SANDSTONES. By T. H. HUXLEY. 14s. 6d.
MONOGRAPH IV. On the CHIMÆROID FISHES of the British Cretaceous Rocks. By E. T. NEWTON. 5s.
The VERTEBRATA of the FOREST BED SERIES of NORFOLK and SUFFOLK. By E. T. NEWTON. 7s. 6d.
CATALOGUE of SPECIMENS in the Museum of Practical Geology, illustrative of British Pottery and Porcelain. By Sir H. DE LA BECHE and TRENHAM REEKS. 155 Woodcuts. 2nd Ed. by T. REEKS and F. W. RUDLER. 1s. 6d.; 2s. in boards.
A DESCRIPTIVE GUIDE to the MUSEUM of PRACTICAL GEOLOGY, with Notices of the Geological Survey, the School of Mines, and the Mining Record Office. By ROBERT HUNT and F. W. RUDLER. 6d. (3rd Ed.)
A DESCRIPTIVE CATALOGUE of the ROCK SPECIMENS in the MUSEUM of PRACTICAL GEOLOGY, By A. C. RAMSAY, H. W. BRISTOW, H. BAUERMAN, and A. GEIKIE. 1s. (3rd Edit.) (*Out of print*.) 4th Ed. in progress.
CATALOGUE of the FOSSILS in the MUSEUM of PRACTICAL GEOLOGY:
  CAMBRIAN and SILURIAN, 2s. 6d.; CRETACEOUS, 2s. 9d.; TERTIARY and POST-TERTIARY, 1s. 8d.

SHEET MEMOIRS OF THE GEOLOGICAL SURVEY.

Those marked (O.P.) are Out of Print.

Sheet	Title
4	FOLKESTONE and RYE. By F. DREW. 1s.
7	PARTS of MIDDLESEX, &c. By W. WHITAKER. 2s. (O.P.)
10	TERTIARY FLUVIO-MARINE FORMATION of the ISLE of WIGHT. By EDWARD FORBES. 5s.
10	The ISLE OF WIGHT. By H. W. BRISTOW. 6s. (O.P.)
12	S. BERKSHIRE and N. HAMPSHIRE. By H. W. BRISTOW and W. WHITAKER. 3s. (O.P.)
13	PARTS of OXFORDSHIRE and BERKSHIRE. By E. HULL and W. WHITAKER. 3s. (O.P.)
34	PARTS of WILTS. and GLOUCESTERSHIRE. By A. C. RAMSAY, W. T. AVELINE, and E. HULL. 8d.
44	CHELTENHAM. By E. HULL. 2s. 6d.
45	BANBURY, WOODSTOCK, and BUCKINGHAM. By A. H. GREEN. 2s.
45 SW	WOODSTOCK. By E. HULL. 1s.
47	N.W. ESSEX & N.E. HERTS. By W. WHITAKER, W. H. PENNING, W. H. DALTON, & F. J. BENNETT. 3s. 6d.
48 SW	COLCHESTER. By W. H. DALTON. 1s. 6d.
48 SE	EASTERN END of ESSEX (WALTON NAZE and HARWICH). By W. WHITAKER. 9d.
48 NW, NE	IPSWICH, HADLEIGH, and FELIXSTOW. By W. WHITAKER, W. H. DALTON, and F. J. BENNETT. 2s.
49 S, 50 SE	ALDBOROUGH, &c. By W. H. DALTON. Edited, with additions, by W. WHITAKER. 1s.
50 SW	STOWMARKET. By W. WHITAKER, F. J. BENNETT, and J. H. BLAKE. 1s.
50 NW	DISS, EYE, &c. By F. J. BENNETT. 2s.
51 SW	CAMBRIDGE. By W. H. PENNING and A. J. JUKES-BROWNE. 4s. 6d.
51 SE	BURY ST. EDMUNDS and NEWMARKET. By F. J. BENNETT, J. H. BLAKE, and W. WHITAKER. 1s. 6d.
53 SE	PART of NORTHAMPTONSHIRE. By W. T. AVELINE and RICHARD TRENCH. 8d.
53 NE	PARTS of NORTHAMPTONSHIRE and WARWICKSHIRE. By W. T. AVELINE. 8d. (O.P.)
63 SE	PART of LEICESTERSHIRE. By W. TALBOT AVELINE, and H. H. HOWELL. 8d. (O.P.)
64	RUTLAND, &c. By J. W. JUDD. 12s. 6d.
66 NE, SE	NORWICH. By H. B. WOODWARD. 7s.
66 SW	ATTLEBOROUGH. By F. J. BENNETT. 1s. 6d.
68 E	CROMER. By C. REID. 6s.
68 NW, SW	FAKENHAM, WELLS, &c. By H. B. WOODWARD. 2s.
70	SW LINCOLNSHIRE, &c. By A. J. JUKES-BROWNE and W. H. DALTON. 4s.
71 NE	NOTTINGHAM. By W. T. AVELINE. (2nd Ed.) 1s.
79 NW	RHYL, ABERGELE, and COLWYN. By A. STRAHAN. (Notes by R. H. TIDDEMAN.) 1s. 6d.
80 NW	PRESCOT, LANCASHIRE. By E. HULL (3rd Ed.). With additions by A. STRAHAN.) 3s.
80 NE	ALTRINCHAM, CHESHIRE. By E. HULL. 8d. (O.P.)
80 SW	CHESTER. By A. STRAHAN. 2s.
81 NW, SW	STOCKPORT, MACCLESFIELD, CONGLETON, & LEEK. By E. HULL and A. H. GREEN. 4s.
82 SE	PARTS of NOTTINGHAMSHIRE and DERBYSHIRE. By W. T. AVELINE. (2nd Ed.) 6d.
82 NE	PARTS of NOTTINGHAMSHIRE, YORKSHIRE, and DERBYSHIRE. By W. T. AVELINE. 8d.
84	EAST LINCOLNSHIRE. By A. J. JUKES-BROWNE.
87 NW	PARTS of NOTTS, YORKSHIRE, and DERBYSHIRE. (2nd Ed.) By W. T. AVELINE. 6d.
87 SW	BARNSLEY. By A. H. GREEN. 9d.
88 SW	OLDHAM. By E. HULL. 2s.
88 SE	PART of the YORKSHIRE COAL-FIELD. By A. H. GREEN, J. R. DAKYNS, and J. C. WARD. 1s.
88 NE	DEWSBURY, HUDDERSFIELD, and HALIFAX. By A. H. GREEN, J. R. DAKYNS, J. C. WARD, and R. RUSSELL. 6d.
89 SE	BOLTON, LANCASHIRE. By E. HULL. 2s.
89 SW	WIGAN. By EDWARD HULL (2nd Ed.) 1s. (O.P.)
90 SE	The COUNTRY between LIVERPOOL and SOUTHPORT. By C. E. DE RANCE. 3d. (O.P.)
90 NE	SOUTHPORT, LYTHAM, and SOUTH SHORE. By C. E. DE RANCE. 6d.
91 SW	The COUNTRY between BLACKPOOL and FLEETWOOD. By C. E. DE RANCE. 6d.
91 NW	SOUTHERN PART of the FURNESS DISTRICT in N. LANCASHIRE. By W. T. AVELINE. 6d.
92 SE	BRADFORD and SKIPTON. By J. R. DAKYNS, C. FOX-STRANGWAYS, R. RUSSELL, and W. H. DALTON. 6d.
93 NW	NORTH and EAST of HARROGATE. By C. FOX-STRANGWAYS. 1s.
93 NE	The COUNTRY between YORK and MALTON. By C. FOX-STRANGWAYS. 1s. 6d.
93 SW	CARBONIFEROUS ROCKS N. and E. of LEEDS, and the PERMIAN and TRIASSIC ROCKS about TADCASTER. By W. T. AVELINE, A. H. GREEN, J. R. DAKYNS, J. C. WARD, and R. RUSSELL. 6d. (O.P.)

SHEET MEMOIRS OF THE GEOLOGICAL SURVEY—continued.

93 SE, 94 SW The COUNTRY between YORK and HULL. By J. R. DAKYNS, C. FOX-STRANGWAYS, and A. G. CAMERON.
94 NW - DRIFFIELD By J. R. DAKYNS and C. FOX-STRANGWAYS.
94 NE - BRIDLINGTON BAY. By J. R. DAKYNS and C. FOX-STRANGWAYS. 1s.
95 SW, SE - SCARBOROUGH and FLAMBOROUGH HEAD. By C. FOX-STRANGWAYS. 1s.
95 NW - WHITBY and SCARBOROUGH. By C. FOX-STRANGWAYS and G. BARROW. 1s. 6d.
96 SE - NEW MALTON, PICKERING, and HELMSLEY. By C. FOX-STRANGWAYS. 1s.
96 NE - ESKDALE, ROSEDALE, &c. By C. FOX-STRANGWAYS, C. REID and G. BARROW. 1s. 6d.
96 NW, SW NORTHALLERTON and THIRSK. By C. FOX-STRANGWAYS, A. G. CAMERON, and G. BARROW. 1s. 6d.
98 SE - KIRKBY LONSDALE and KENDAL. By W. T. AVELINE, T. MCK. HUGHES, and R. H. TIDDEMAN. 2s.
98 NE - KENDAL, WINDERMERE, SEDBERGH, & TEBAY. By W. T. AVELINE & T. MCK. HUGHES. 6d. (O.P.)
101 SE - NORTHERN PART of the ENGLISH LAKE DISTRICT. By J. C. WARD. 9s.
104 SW, SE NORTH CLEVELAND. By G. BARROW.
108 SE - OTTERBURN and ELSDON. HUGH MILLER. (Notes by C. T. CLOUGH.)

THE MINERAL DISTRICTS OF ENGLAND AND WALES ARE ILLUSTRATED BY THE FOLLOWING PUBLISHED MAPS OF THE GEOLOGICAL SURVEY.

COAL-FIELDS OF ENGLAND AND WALES.

Scale, one inch to a mile.

Anglesey, 78 (SW).
Bristol and Somerset, 19, 35.
Coalbrook Dale, 61 (NE & SE).
Clee H. 11, 53 (NE, NW).
Flintshire and Denbighshire, 74 (NE & SE), 79 (NE, SE).
Derby and Yorkshire, 71 (NW, NE, & SE), 82 (NW & SW), 81 (NE), 87 (NE, SE), 88 (SW).
Forest of Dean, 43 (SE & SW).
Forest of Wyre, 61 (SE), 55 (NE).
Lancashire, 80 (NW), 81 (NW), 89, 88 (SW, NW).
Leicestershire, 71 (SW), 63 (NW).
Northumberland & Durham, 103, 105, 106 (SE), 109 (SW, SE).
N. Staffordshire, 72 (SW), 73 (NE), 80 (SE), 81 (SW).
S. Staffordshire, 54 (NW), 62 (SW).
Shrewsbury, 60 (NE), 61 (NW & SW).
South Wales, 36, 37, 38, 40, 41, 42 (SE, SW).
Warwickshire, 62 (NE SE), 63 (NW SW), 54 (NE), 53 (NW).
Yorkshire, 88 (NE, SE), 87 (SW), 92 (SE), 93 (SW).

GEOLOGICAL MAPS.

Scale, six inches to a mile.

The Coal-fields and other mineral districts of the N. of England are published on a scale of six inches to a mile, at 4s. to 6s. each. MS. Coloured Copies of other six-inch maps, not intended for publication, are deposited for reference in the Geological Survey Office, 28, Jermyn Street, London.

Lancashire.

Sheet.	Sheet.	Sheet.
15. Ireleth.	75. Todmorden.	97. Oldham.
16. Ulverstone.	77. Chorley.	100. Knowsley.
17. Cartmel.	78. Bolton-le-Moors.	101. Billinge.
22. Aldingham.	79. Entwistle.	102. Leigh, Lowton.
47. Clitheroe.	80. Tottington.	103. Ashley, Eccles.
48. Colne.	81. Wardle.	104. Manchester, Salford.
49. Laneshaw Br.	84. Ormskirk.	105. Ashton-under-Lyne.
55. Whalley.	85. Standish.	
56. Haggate.	86. Adlington.	106. Liverpool.
57. Winewall.	87. Bolton-le-Moors.	107. Prescott.
61. Preston.	88. Bury, Heywood.	108. St. Helen's.
62. Balderstone.	89. Rochdale, &c.	109. Winwick.
63. Accrington.	92. Bickerstaffe.	111. Cheadle.
64. Burnley.	93. Wigan.	112. Stockport.
65. Stiperden Moor.	94. West Houghton.	113. Part of Liverpool.
39. Layland.	95. Radcliffe.	
70. Blackburn.	96. Middleton, Prestwich.	
71. Haslingden.		
72. Cliviger, Bacup.		

Durham.

1. Ryton.	6. Winlaton.	11. Ebchester.
2. Gateshead.	7. Washington.	12. Tantoby.
3. Jarrow.	8. Sunderland.	13. Chester-le-St.
4. S. Shields.	9. ———	16. Hunstanworth.
5. Greenside.	10. Edmondbyers.	17. Waskerley.

Sheet.	Sheet.	Sheet.
18. Muggleswick.	25. Wolsingham.	38. Maize Beck.
19. Lanchester.	26. Brancepeth.	41. Cockfield.
20. Hetton-le-Hole.	30. Benny Seat.	42. Bp. Auckland.
22. Wear Head.	32. White Kirkley.	46. Hawksley Hill Ho.
23. Eastgate.	33. Hamsterley.	52. Barnard Castle.
24. Stanhope.	34. Whitworth.	53. Winston.

Northumberland.

44. Rothbury.	80. Cramlington.	98. Walker.
45. Longframlington.	81. Earsdon.	101. Whitfield.
46. Broomhill.	82. NE. of Gilsland.	102. Allendale Town.
47. Coquet Island.	83. Coadley Gate.	103. Slaley.
54. Longhorsley.	87. Heddon.	105. Newlands.
55. Ulgham.	88. Long Benton.	106. Blackpool Br.
56. Druridge Bay.	89. Tynemouth.	107. Allendale.
63. Netherwitton.	91. Greenhead.	108. Blanchland.
64. Morpeth.	92. Haltwhistle.	109. Shotleyfield.
65. Newbiggin.	93. Haydon Bridge.	110. Wellhope.
72. Bedlington.	94. Hexham.	111. Allenheads.
73. Blyth.	95. Corbridge.	112.
	96. Horsley.	
	97. Newcastle.	

Cumberland.

2. Tees Head.	65. Dockraye.	74. Wastwater.
6. Dufton Fell.	69. Buttermere.	75. Stonethwaite Fell.
	70. Grange.	
55. Searness.	71. Helvellyn.	
56. Skiddaw.		
63. Thackthwaite.		
64. Keswick.		

Westmorland.

	12. Patterdale.	25. Grasmere.
	18. Near Grasmere.	38. Kendal.

Yorkshire.

7. Redcar.	116. Conistone Moor.	260. Honley.
9. ———	133. Kirkby Malham.	261. Kirkburton.
12. Bowes.	184. Dale End.	262. Darton.
13. Wycliffe.	185. Kildwick.	263. Hemsworth.
20. Lythe.	200. Keighley.	264. Campsall.
24. Kirkby Ravensworth.	201. Bingley.	272. Holmfirth.
25. Aldborough.	202. Calverley.	273. Penistone.
32. Whitby.	203. Seacroft.	274. Barnsley.
33. ———	204. Aberford.	275. Darfield.
38. Marske.	215. Peeke Well.	276. Brodsworth.
39. Richmond.	216. Bradford.	281. Langsell.
46. ———	217. Calverley.	282. Wortley.
47. Robin Hood's Bay.	218. Leeds.	283. Wath upon Dearne.
53. Downholme.	219. Kippax.	284. Conisborough.
68. Leybourne.	231. Halifax.	287. Low Bradford.
82. Kidstones.	232. Birstal.	288. Ecclesfield.
84. E. Witton.	233. East Ardsley.	289. Rotherham.
97. Foxup.	234. Castleford.	290. Braithwell.
98. Kirk Gill.	246. Huddersfield.	293. Hallam Moors.
99. Haden Carr.	247. Dewsbury.	295. Handsworth.
100. Lofthouse.	248. Wakefield.	296. Laughton-en-le-Morthen.
115. Arncliffe.	249. Pontefract.	299. ———
	250. Darrington.	300. Harthill.

MINERAL STATISTICS.

Embracing the produce of Coals, Metallic Ores, and other Minerals. By R. HUNT. From 1853 to 1857, inclusive, 1s. 6d. each. 1858, Part I., 1s. 6d.; Part II., 5s. 1859, 1s. 6d. 1860, 3s. 6d. 1861, 2s.; and Appendix, 1s. 1862, 2s. 6d. 1863, 2s. 6d. 1864, 2s. 1865, 2s. 6d. 1866 to 1881, 2s. each.

(These Statistics are now published by the Home Office, as parts of the Reports of the Inspectors of Mines.)

THE IRON ORES OF GREAT BRITAIN.

Part I. The North and North Midland Counties of England (Out of print). Part II. South Staffordshire. Price 1s.
Part III. South Wales. Price 1s. 3d. Part IV. The Shropshire Coal-field and North Staffordshire. 1s. 3d.

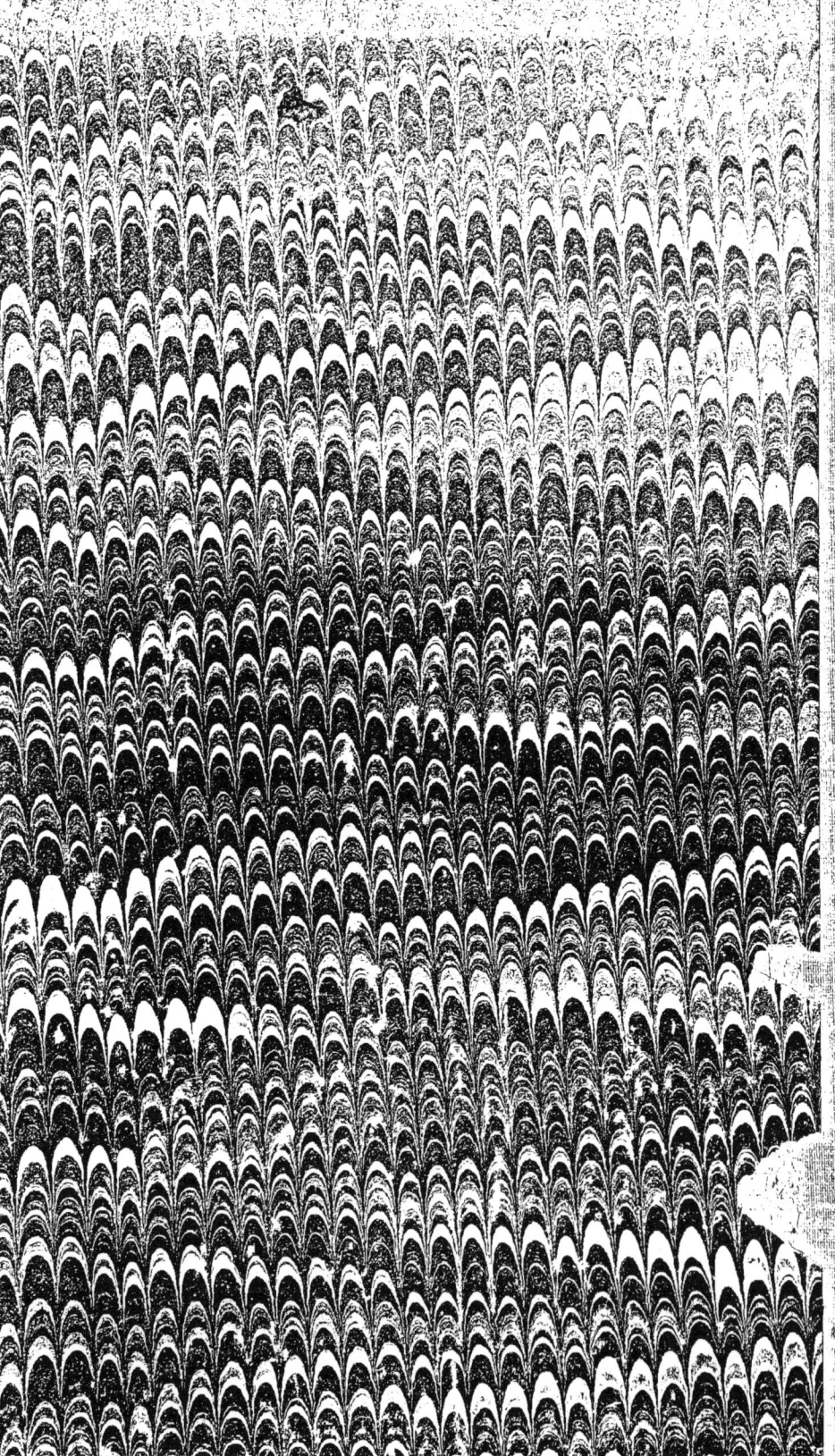

This volume from the
Cornell University Library's
print collections was scanned on an
APT BookScan and converted
to JPEG 2000 format
by Kirtas Technologies, Inc.,
Victor, New York.
Color images scanned as 300 dpi
(uninterpolated), 24 bit image capture
and grayscale/bitonal scanned
at 300 dpi 24 bit color images
and converted to 300 dpi
(uninterpolated), 8 bit image capture.
All titles scanned cover to
cover and pages may include
marks, notations and other
marginalia present in the
original volume.

The original volume was digitized
with the generous support of the
Microsoft Corporation
in cooperation with the
Cornell University Library.

Cover design by Lou Robinson,
Nightwood Design.

Printed in Great Britain
by Amazon